Also by Richard Wiseman

The Luck Factor
Did You Spot the Gorilla?
Quirkology
59 Seconds
Paranormality

THE
AS IF
PRINCIPLE

*The Radically New Approach
to Changing Your Life*

Richard Wiseman

*f*P

Free Press
A Division of Simon & Schuster, Inc.
1230 Avenue of the Americas
New York, NY 10020

Copyright © 2012 by Spin Solutions, Ltd.

First published in Great Britain in 2012 by Macmillan UK

All rights reserved, including the right to reproduce this book or portions thereof in any form whatsoever. For information address Free Press Subsidiary Rights Department, 1230 Avenue of the Americas, New York, NY 10020

First Free Press hardcover edition January 2013

FREE PRESS and colophon are trademarks of Simon & Schuster, Inc.

For information about special discounts for bulk purchases, please contact Simon & Schuster Special Sales at 1-866-506-1949 or business@simonandschuster.com.

The Simon & Schuster Speakers Bureau can bring authors to your live event. For more information or to book an event contact the Simon & Schuster Speakers Bureau at 1-866-248-3049 or visit our website at www.simonspeakers.com.

Designed by Carla Jayne Jones

Manufactured in the United States of America

10 9 8 7 6 5 4 3 2 1

Library of Congress Cataloging-in-Publication Data

Wiseman, Richard
The as if principle : the radically new approach to changing your life / by Richard Wiseman.
p. cm.
Previously published under title: Rip it up.
Includes bibliographical references and index.
1. Self-actualization (Psychology) 2. Interpersonal relations. I. Title.
BF637.S4W576 2012
158—dc23   2012025667

ISBN 978–1–4516–7505–4
ISBN 978–1–4516–7507–8 (ebook)

To Ronald and Brenda

If you want a quality, act as if you already have it.
—William James, 1884

CONTENTS

A BRIEF INTRODUCTION   xi
It's time for positive action

CHAPTER 1: HOW TO BE HAPPY   1
Where we meet that adorable genius William James, turn the world upside-down, learn how to create good cheer at will, and visit the fun factory

CHAPTER 2: ATTRACTION AND RELATIONSHIPS   33
Where we unpack the mysteries of the human heart, discover the power of footsie, invent a new type of speed dating, and learn how to live happily ever after

CHAPTER 3: MENTAL HEALTH   73
Where we meet the "Napoleon of neuroses," find out why watching sports is bad for your health, and discover how best to deal with phobias, anxiety, and depression

CHAPTER 4: WILLPOWER   115
Where we learn why rewards punish and discover how to motivate others, beat procrastination, stop smoking, and lose weight

CONTENTS

CHAPTER 5: PERSUASION   145
   Where we explore the problems of changing people's minds,
   find out what really manipulates the masses, and discover how
   cooperation can shape society

CHAPTER 6: CREATING A NEW YOU   187
   Where we learn how to feel more confident, change our
   personality, and slow the effects of aging

CONCLUSION   233
   Where we hypnotize a woman, saw a brain in half, and discover
   why you really are of two minds about everything

Appendix   243
Acknowledgments   247
Notes   249
Index   267

A Brief Introduction

It's time for positive action

Self-help gurus and business coaches preach the same simple mantra: if you want to improve your life, you need to change how you think. Force yourself to have positive thoughts, and you will become happier. Visualize your dream self, and you will enjoy increased success. Think like a millionaire, and you will magically grow rich. In principle, this idea sounds perfectly reasonable. However, in practice, the approach often proves surprisingly ineffective, with research showing that people struggle to continually think happy thoughts, that employees remain unmoved by imagining their perfect selves, and that those dreaming of endless wealth fail to make millions.

Over a century ago, the brilliant Victorian philosopher William James proposed a radically different approach to change. Since then, researchers from across the world have carried out hundreds of experiments into James's theory and discovered that it applies to almost every aspect of people's lives. Perhaps most important of all, the work has given rise to a series of easy and effective exercises that can help people feel happier, avoid anxiety and worry, fall in love and live happily ever after, stay slim, increase their willpower and confidence, and even slow the effects of aging. This research has been presented at countless scientific conferences and published in academic journals, but it has rarely made its way into the public domain.

In my previous book, *59 Seconds,* I described a handful of these exercises. *The As If Principle* builds on this work by presenting the first accessible and comprehensive guide to James's radical theory. It reveals how everything you currently believe about your mind is wrong, shows that change does not have to be challenging, and describes a series of easy-to-implement techniques designed to improve several different areas of your everyday life.

Throughout the book you are going to be presented with interactive exercises that involve you changing your behavior. You might be tempted to read about these ideas but not actually put them into practice. This attitude is perhaps understandable (after all, behaviors quickly become ingrained in our minds and soon start to feel like old friends), but please resist the urge. This book is all about the impact that your actions have on your mind, and the exercises encourage you to actually experience these phenomena rather than just read about them.

You are about to encounter a new approach to change. An approach that is grounded in science, overturns conventional thinking, and provides a basis for the easiest, quickest and most effective ways of changing your life.

So sit up straight and take a deep breath. Forget all about positive thinking. It's time for positive action.

THE
AS IF
PRINCIPLE

Chapter 1

How to Be Happy

Where we meet that adorable genius William James, turn the world upside down, learn how to create good cheer at will, and visit the fun factory

"In the beginning was the deed."

—*Faust*, Johann Wolfgang von Goethe

I. The Simple Idea That Changes Everything
II. Testing a Theory
III. The Value of Fun

I. The Simple Idea That Changes Everything

The world's first laboratory-based psychology experiment was carried out by the German psychologist Professor Wilhelm Wundt in 1879. This historic study was conducted in a small room at the University of Leipzig and reveals all you need to know about how Victorian scientists approached the human mind.

Wundt could have celebrated the birth of experimental psychology by investigating any fascinating topic of his choosing. Perhaps why people fall in love, believe in God, or sometimes feel the need to kill one another. Instead, the "humourless and indefatigable" Wundt chose to conduct a strange, even bizarre, experiment involving a small brass ball.[1]

Wundt and two of his students gathered around a small table and connected a timer, a switch, and a carefully designed metal stand. A brass ball was then balanced on the stand, and one of Wundt's students placed his hand a few millimeters above the switch. Seconds later, the ball was automatically released from the stand, and the timer sprang into action. The student slammed his hand down on the switch the moment that he heard the brass ball hit the table, which immediately halted the timer. By carefully recording the reading shown on the timer in his notebook, Wundt created psychology's first data point.

It would be nice to think that after a day or so of ball dropping, Wundt would have closed his notebook, reported his findings, and moved on to something more interesting. Nice, but wrong. In fact, Wundt spent the next few years of his life observing hundreds of people responding to the same test. In the same way that physicists were trying to identify the fundamental nature of matter, so Wundt and his team were attempting to discover the fundamental building blocks of consciousness. Some of the participants were asked to press the switch the moment they heard the ball hit the table, while others were told to react when they became fully aware of the sound. In the first scenario, observers were asked to concentrate their attention on the ball, and in the second, they were asked to focus more on their own thoughts. When the tasks were performed properly, Wundt believed that the first reaction would represent a simple reflex, whereas the second was more of a conscious decision. Perhaps not surprisingly, many participants initially struggled to recognize the alleged subtle difference between the two conditions, and so were required to complete more than ten thousand trials before moving on to the experiment proper.

After carefully wading through the resulting mass of ball-dropping data, Wundt concluded that the reflexive response took an average of one-tenth of a second and left the participants with a very weak mental record of the sound of the ball. In contrast, consciously hearing the sound produced an average reaction time of two-tenths of a second and resulted in a far clearer experience of the ball's impact.

Having solved the mystery of the reflexive response, Wundt devoted the rest of his career to carrying out hundreds of similar studies. His approach proved surprisingly influential, and almost every other nineteenth-century academic dabbling in matters of the mind followed in his footsteps. In psychology laboratories across Europe, researchers could hardly hear themselves think for the sound of brass balls dropping onto tables.

In America, a young philosopher and psychologist named William James was having none of it.

William James, born in 1842 in New York City, was a most remarkable man. His father was an independently wealthy, well-connected, eccentric, one-legged religious philosopher who devoted himself to educating his five children.[2] As a result, much of James's childhood was spent receiving private tutoring, visiting Europe's leading museums and art galleries, and rubbing shoulders with the likes of Henry Thoreau, Alfred Tennyson, and Horace Greeley. James's older brother, Henry, would go on to find fame as a novelist and his sister, Alice, as a diarist.

Initially trained in painting, James abandoned the arts in his twenties and enrolled to study chemistry and anatomy at Harvard Medical School. In 1872 family friend and Harvard president Charles Eliot recruited James to teach courses in vertebrate physiology. James soon found himself drawn to the mysteries of the human psyche and in 1875 put together America's first psychology course, later remarking that "the first lecture in psychology that I ever heard was the first one I gave."

Appalled by what he saw as the triviality of Wundt's work, James firmly believed that psychological research should be relevant to people's lives. Turning his back on brass balls and reaction times, James instead focused his attention on a series of far more interesting and pragmatic issues, including whether it was right to believe in God, what made life worth living, and if free will actually exists.

Wundt and James didn't differ just in terms of their approach to the human mind.

Wundt was formal and stuffy, his lectures serious and solemn, and his writing dull and turgid. James was informal and unpretentious, often walking around campus sporting "a silk hat, cane, frock coat and red-checked trousers." He frequently peppered his talks with jokes and light-hearted asides to the extent that his students often felt the need to ask him to be more serious, and he produced accessible and often amusing prose ("As long as one poor cockroach feels the pangs of unrequited love, this world is not a moral world").

James and Wundt also developed completely different ways of working. Wundt recruited a large team of students to conduct his carefully controlled studies. On their first day in Wundt's laboratory, each new intake of students would be lined up, and Wundt would move down the line handing each of them a description of the research that they were required to conduct. Once the work was completed, he acted as judge and jury, and any students reporting results that failed to support their master's theories ran the risk of failing.[3] In contrast, James loved to encourage free thought, loathed the idea of imprinting his ideas on students, and once complained that he had just seen a fellow academic "applying the last coat of varnish to his pupil."

The two great thinkers did little to hide their animosity for each other. James developed a poetic turn of phrase, causing some commentators to note that he wrote psychology papers like a novelist, while his brother, Henry, penned novels like a psychologist. Wundt, however, remained unimpressed, and once when he was asked to comment on James's writings, he replied, "It is beautiful, but it is not psychology." In reply, James complained about Wundt's altering his theories from one book to the next, noting, "Unfortunately he will never have a Waterloo . . . cut him up like a worm and each fragment crawls . . . you can't kill him."

Despite being vastly outnumbered by Wundt's army of supporters, James stood his ground. While almost every psychologist in Europe was obsessively carrying out increasingly esoteric variations of Wundt's classic ball-dropping experiment, James continued to stroll around Harvard in his red-checked trousers encouraging his students to think about the meaning of life.

James's persistence paid off. Open any recent psychology textbook, and you will find nary a passing reference to Wundt or his brass balls. In contrast, James's ideas are still widely cited, and he is seen as the founding father of modern psychology. First published in 1890, James's two-volume magnum opus, *Principles of Psychology*, was recently described by one leading historian as "the most literate, most provocative, and at the same time the most intelligent book on psychology that has ever appeared,"[4] and both volumes are still considered required reading for students of behavioral science today. Harvard's psychology department named its building after James, and each year the Association for Psychological Science gives its William James Fellow Award to the academic judged to have made the most significant intellectual contribution to psychology.

James was perhaps at his best when he found mystery and substance in phenomena that most people tended to take for granted. In 1892 he reflected on the importance of this approach to understanding the human mind, and provided a few examples of the types of phenomena that had recently caught his attention:

> Why do we smile, when pleased, and not scowl? Why are we unable to talk to a crowd as we talk to a single friend? Why does a particular maiden turn our wits so upside-down? The common man can only say, "*Of course* we smile, *of course* our heart palpitates at the sight of the crowd, *of course* we love the maiden, that beautiful soul clad in that perfect form, so palpably and flagrantly made for all eternity to be loved!"[5]

It was exactly this kind of thinking that led James to create his most controversial theory and turn our understanding of the human mind on its head.

Toward the end of the 1880s, James turned his attention to the relationship between emotion and behavior. To the uninitiated, this may seem a strange choice of topic for a world-renowned philosopher and psychologist.

Common sense suggests that certain events and thoughts cause you to feel emotional and that this in turn affects your behavior. So, for example, you might find yourself walking along an unexpectedly dark street late at night, or being called into your boss's office and awarded a pay raise, or suddenly remembering a time you were five years old and fell down the stairs. These stimuli cause you to experience certain emotions. Perhaps the dark street makes you feel anxious, the pay raise makes you feel happy, and the memory of falling down the stairs makes you feel upset. These emotions then affect your behavior. Feeling afraid may make you sweat, feeling happy may make you smile, and feeling upset may make you cry. Seen from this perspective, the link between how you feel and the way you act is as straightforward as it is unsurprising. Mystery solved, case closed.

Behavior and Emotion

Common sense suggests that emotions cause behavior:

Feel anxious	— Sweat
Feel happy	— Smile
Feel sad	— Cry

However, James's previous experience with seemingly straightforward psychological phenomena made him well aware that conventional wisdom can often be deeply misleading. Take, for example, James's work on memory. For years armchair philosophers had suggested that memory operated much like a muscle, believing that the more you used it, the stronger it became. James wondered whether this was really accurate.[6] To find out, he spent eight days timing himself as he memorized 158 lines of the Victor Hugo poem "Satyr" and discovered that the task took him an average of fifty seconds per line. Then, to exercise his memory muscle further, he devoted twenty minutes each day for the following thirty days memorizing the entire first book of Milton's *Paradise Lost*. If the theory of the more you use it, the stronger it gets was correct, James hypothesized that he should be able to return to "Satyr" and learn the next 158 lines in less time than before. In fact, when he tried to learn another section of the poem, he discovered that it took him longer than before. The memory-as-muscle hypothesis was wrong.

James wanted to explore whether there was an alternative to the commonsense theory of emotion, and began his intellectual quest by thinking about how we go about deciding how other people feel.

Look at the photograph in Figure 1 below and try to imagine how the two people in the photograph are feeling. Now do the same with the people in Figure 2. Most people find this exercise easy. Almost everyone assumes that the people in the first photograph are probably having a good time and are likely to be experiencing happiness with just a hint of attraction. The second photograph elicits a quite different reaction, with most people concluding that everyone in the group is probably concerned and anxious and that at least one of them appears to be in need of a comfort break.

This simple exercise is based on an experiment first conducted by the legendary naturalist Charles Darwin in the mid-1800s. Darwin published twenty-two books in his lifetime, including the ground-breaking *On the Origin of Species by Means of Natural Selection, or the Preservation of Favoured Races in the Struggle for Life* and his lesser-known tome, *The Formation of Vegetable Mould Through the Action of Worms, with Observations on Their Habits.* In 1872 Darwin published a seminal text on emotion, *The Expression of the Emotions in Man and Animals,* and described carrying out the first psychological study of emotions.[7]

A French physician named Guillaume-Benjamin-Amand Duchenne had previously applied painful electrical shocks to the muscles in a volunteer's face in order to study facial anatomy. When Darwin saw photographs of Duchenne's work, he was struck by how easily he associated emotions with the volunteer's expressions. Intrigued, Darwin showed some of the photographs to his friends and asked them to say which emotion the volunteer appeared to be feeling. Darwin's friends also reliably and effortlessly associated certain expressions with particular emotions, proving that the ability to know how others are feeling based on their facial expressions is somehow built into our brains.

James read about Darwin's experiment and used it as a basis for his new theory about emotion. Darwin had shown that people are extremely skilled at knowing how another person feels from facial expression. James wondered whether exactly the same mechanism might also account for how they themselves experience emotions. He suggested that in the same way that you look at other people's expressions and work out how they feel, so you might monitor your own expressions and then decide what emotion you should experience.

James originally proposed that any emotion is entirely the result of people observing their own behavior. Seen from this perspective, people smile not because they are happy but rather always feel happy because they are smiling (or, to use James's more poetic way of explaining his radical hypothesis, "You do not run from a bear because you are afraid of it, but rather become afraid of the bear because you run from it"). James makes a clear distinction between our body's instinctive physical behavior in the face of stimulus—whether pulling our hands away from a flame, smiling at a joke, or instantly starting to turn on our heel at the sight of an angry bear—and how our brain observes that movement and, a split second later, produces an emotion. You see the bear, your body behaves by starting to run, and your brain decides, "I'm afraid." More modern versions of James's theory view the relationship between emotion and behavior as a two-way street, suggesting, for example, that people smile because they are happy and also become happier when they smile.

James never formally tested his theory because he found experimentation both boring and intellectually unrewarding ("The thought of brass-instrument and algebraic-formula psychology fills me with horror"). He was, however, a passionate pragmatist and lost no time exploring the potential practical implications of his idea.

The notion of behavior causing emotion suggests that people should be able to create any feeling they desire simply by acting as if they are experiencing that emotion. Or as James famously put it, "If you want a quality, act *as if* you already have it." I refer to this simple but powerful proposition as the As If principle (see the diagram).

Behavior and Emotion

Common sense suggests that the chain of causation is:
   You feel happy — You smile
   You feel afraid — You run away
The As If theory suggests that the opposite is also true:
   You smile   — You feel happy
   You run away   — You feel afraid

This aspect of James's theory clearly energized him more than any other. In one public talk, he described the potential power of the idea as "bottled lightning" and enthusiastically noted, "The sovereign voluntary path to cheerfulness . . . is to sit up cheerfully, to look round cheerfully, and to act and speak as if cheerfulness were already there. . . . To wrestle with a bad feeling only pins our attention on it, and keeps it still fastened in the mind."[8]

James's theory met with a critical reaction from some of his peers. Wilhelm Wundt roundly condemned the idea, labeling it a *"psychologischen Scheinerklarungen"* and presented his own account of emotion, which suggested that feelings were a *Gefühl,* defined as "an unanalyzable and simple process corresponding in the sphere of Gemuth to sensation in the sphere of intellection" (good to get that sorted out). James defended his position, but the theory proved too radical for many of his more conventional colleagues and was quickly relegated to the filing drawer marked "Years ahead of its time."

And there it lay for more than sixty years.

II. Testing a Theory

In the late 1960s a young academic named James Laird was studying for his doctorate in clinical psychology at the University of Rochester.[9] During a training session there, Laird was asked to interview a patient while his supervisor watched through a one-way mirror. At one point in the interview, a rather unusual smile spread across the patient's face. Laird was intrigued by the smile and wondered how the patient had felt when he had produced the rather odd facial expression.

As Laird drove home from the interview, he replayed the session in his mind and became fixated on the smile. Eventually he forced his own face into the same expression to discover how it felt. Laird was amazed to discover that the smile instantly made him feel happier. Intrigued, he tried frowning and suddenly felt sad. Those few rather strange moments during his drive home changed the entire course of his career. When he reached his house that night, Laird went straight to his bookshelf where he searched for information about the psychology of emotion. By chance, the first book he picked up was William James's *Principles of Psychology*.

Laird read about James's long-lost theory and realized that it might explain why smiling in his car had made him feel happier. He was also amazed to discover that the theory had been confined to the history books and had never been properly tested. In order to do so, Laird invited volunteers into his laboratory, asked them to smile or frown, and then report how they felt. According to James, those who had been smiling should feel significantly happier than those who had forced a frown on their faces.

However, worried that volunteers might be tempted to tell him what he wanted to hear, Laird wanted to find a way of making people smile or frown while concealing the true nature of the experiment. Eventually he hit on a clever cover story. He told the volunteers that they would be participating in a study examining the electrical activity in their facial muscles and placed electrodes between the participant's eyebrows, at the corners of their mouth, and at the corners of their jaw. He then explained that changes in their emotion could affect the experiment, and so to rule out this possible source of error, they would be asked to report their emotions as the experiment proceeded.

The electrodes were fake, but the clever cover story allowed Laird secretly to manipulate his volunteers' faces into a smile or frown. To create an angry expression, the participants were asked to draw down and pull together the two electrodes between the eyebrows and contract the ones on the jaw by clenching their teeth. For the happy expression, they were asked to draw back the electrodes at the corners of the mouth toward the back of the face.

After they had contorted their face into the required position, participants were presented with a checklist containing a list of emotions (such as aggression, anxiety, joy, and remorse) and asked to rate the degree to which they were experiencing each. The results were remarkable. Exactly as James had predicted at the turn of the twentieth century, the participants felt significantly happier when they forced their face into smiles and significantly angrier when they were frowning.

After the study Laird interviewed his participants and asked them if they knew why they had experienced various emotions during the study. Only a handful put their emotional state down to their manipulated expressions, with the rest at a loss to explain the shift. In one of the interviews, a participant whose face had been contorted into a frown explained, "I'm not in any angry mood, but I found my thoughts wandering to things that made me angry, which is sort of silly, I guess. I knew I was in an experiment, and I knew I had no reason to feel that way, but I just lost control."

How to Be Happy in an Instant

Around the turn of the last century, the Russian theater director Constantin Stanislavski revolutionized drama by creating method acting. A key part of his approach is to encourage actors to experience genuine emotion on stage by controlling their behavior. This technique, often referred to as the "magic if" ("If I was really experiencing this feeling, how would I behave?"), has been adopted by several famous performers, including Marlon Brando, Warren Beatty, and Robert De Niro.

The same technique has been used in laboratory experiments exploring the As If principle. Let's imagine that you are taking part in a study to test the As If principle. At the start of the study, you would be asked to rate how cheerful you feel on a scale between 1 (how you would feel if you had just fallen down an open manhole) and 10 (how you would feel if you had just seen your worst enemy do exactly the same thing).

Next, you would be asked to smile. However, there is more to acting happy than simply forcing your face into a brief, unfelt smile that finishes in the blink of an eye. Instead, you would be asked to follow these instructions:

1. Sit in front of a mirror.
2. Relax the muscles in your forehead and cheeks, and let your mouth drop slightly open. In scientific circles, the expression that you have on your face right now is referred to as neutral and acts as a blank canvas.
3. Contract the muscles near the corners of your mouth by drawing them back toward your ears. Make the resulting smile as wide as possible and try to ensure

that the movement of the cheeks produces wrinkling around the base of your eyes. Finally, extend your eyebrow muscles slightly upward and hold the resulting expression for about twenty seconds.

4. Let the expression drop from your face and think about how you feel.

Do you feel more cheerful than before you started? What number would you give to this new feeling on the scale of 1 to 10?

Most people report that the exercise has made them feel happier. As predicted by William James more than a century ago, just a few seconds changing your facial expression has a big impact on how you feel.

To boost your level of good cheer, incorporate this type of smiling into your daily routine. Create a fun way of reminding yourself to do this by drawing two self-portraits of yourself wearing a huge smile. Draw one of the portraits on a sheet of letter-size paper and the other on a small piece of paper that is about two inches square. Make the portraits as humorous and happy looking as possible. Finally, place the large portrait somewhere prominent in your home and the smaller one in your wallet or purse, and use them as a cue to help you remember to smile.

To make sure that the remarkable effect was genuine, other scientists set about attempting to replicate Laird's groundbreaking result. Rather than repeatedly placing fake electrodes on people's faces, each laboratory produced its own cover story.

Inspired by photographers who make people smile by getting them to say the word *cheese,* researchers at the University of Michigan had participants repeatedly make an *ee* sound (as in *easy*) to force their faces into a smile or an *eu* sound (as in *yule*) to produce an expression nearer to disgust.[10]

Psychologists at Washington University attached a golf tee to the inner ends of their participant's eyebrows and then asked them to carry out one of two facial contortions.[11] Participants in one group were asked to make the golf tees touch by drawing their eyebrows down and together, thus producing an unhappy facial expression. Those in the other group were asked to ensure that the ends of the golf tees didn't touch, thus creating a more neutral expression.

In perhaps the best known of the studies, researchers in Germany told participants that they were investigating a new way of teaching people who were paralyzed below the neck to write.[12] Half of the participants were asked to support a pencil horizontally between their teeth (forcing their faces into a smile) and the other half were asked to hold the pencil between their lips (pulling their faces into a frown).

Participants repeatedly chanting *ee,* keeping their golf tees apart, or supporting a pencil between their teeth suddenly felt significantly happier. Time and again, the research showed that Laird's results were genuine and that James's theory was correct. Your behavior does influence how you feel, and so, as a result, it is possible to manufacture emotions at will, as the As If principle predicts.

Excited by the results, researchers set out to discover the impact that the principle has on your body and brain.

Body and Brain

Paul Ekman from the University of California has devoted his career to studying facial expressions and emotion. During a long and distinguished career, he has produced the definitive guide to facial expression (a five-hundred-page treatise showing how the forty-three facial muscles combine to produce thousands of expressions), advised law enforcement agencies across the world on the best ways of identifying whether someone is telling the truth from his or her facial expressions, and helped to create the hit television show *Lie to Me.*

17

Toward the start of his career, Ekman was intrigued to hear about the notion that altering people's facial expressions could make them feel either relaxed or angry and wanted to discover how the As If principle affects the body. His remarkable results pay tribute to the power of James's theory.

Ekman invited volunteers into his laboratory, where he attached them to a machine that monitored their heart rate and skin temperature.[13] He then asked each participant to carry out two tasks. The first was designed to make them feel genuine anger and involved thinking about an event in their lives that had made them feel angry, then mentally reliving that event as vividly as possible. For the second one, they produced only the facial expression of anger (eyebrows down and together, raised upper eyelid, lower lip up and lips pressed together). This procedure was then repeated for several emotions, including fear, sadness, happiness, surprise, and disgust.

Not surprisingly, the genuine emotional memories triggered certain patterns in participants' physiology with, for example, fear producing a high heart rate and low skin temperature and happiness resulting in a low heart rate and higher skin temperature. Remarkably, exactly the same physiological pattern emerged when people adopted a facial expression. When they looked fearful, their heart rate rocketed and skin temperature dropped. When they put a smile on their faces, their heart rates fell and their skin temperature increased.

Curious to discover whether this mechanism was hardwired into the human psyche, Ekman and his team journeyed across the world and repeated his study with the inhabitants of a remote island in western Indonesia.[14] The results were identical to those found in the West, suggesting that the As If principle is not a product of Western culture but rather the deep-seated product of our evolutionary past. Ekman's findings showed that behaving as if you are experiencing an emotion does more than influence how you feel; it also has a direct and powerful effect on your body.

More recently, researchers have built on this work by using the latest technology to discover how the As If principle affects the brain. If you were to cut off your head and examine the region of the brain closest to the top of your spine, you would see two almond-shaped pieces of tissue on either side of the spinal chord. These are the amygdala (named after the Latin for "almond"). They form a very small, but very well-connected, part of the brain that plays a key role in almost every aspect of your everyday life. The amygdala is central to emotional experiences, especially fear.

The key role that the almond of horror plays in fear was recently illustrated by scientists studying a remarkable patient referred to as "SM."[15] SM suffers from Urbach-Wiethe disease, a rare genetic disorder that causes the amygdala to degenerate. After interviewing SM, the scientists noticed that she described several incidents in her life when she should have experienced fear, but didn't. In perhaps the most dramatic of these, SM was unfortunate to be attacked in a local park. Her attacker held a knife to her throat and threatened to stab her. SM said that she didn't feel afraid at the time, but instead noticed a nearby church and calmly said, "If you're going to kill me, you're gonna have to go through my God's angels first." Confused, the attacker suddenly let her go.

Intrigued, the scientists set out to scare SM. They took her to an exotic pet store and asked her to handle snakes and spiders. SM showed no reaction and had to be stopped from touching the more dangerous ones. Next they took her to an allegedly haunted house and showed her horror movie clips. Again, nothing—proof that a fully functioning amygdala plays a key role in experiencing fear.

A few years ago, scientists decided to conduct the ultimate test of James's hypothesis by putting participants in a brain scanner and asking them to contort their faces into a fearful expression.[16] Unlike the psychological studies that had been conducted throughout the preceding few decades, the participants did not have to tell the experimenters how they were feeling. Instead, the researchers just looked directly inside participants' brains, saw a highly active amygdala, and could conclude that the participants were indeed experiencing genuine fear. In doing so, the researchers obtained the ultimate proof that behaving As If directly influences your brain.

The As If principle has been used to manufacture happiness in laboratories across the world and has the power to have an instant impact on people's bodies and brains. But does the effect work in the real world? Could it even be employed to cheer up an entire nation? It was time to find out.

The Science of Happiness Project

I have conducted several mass participation experiments during my career. These studies have involved tens of thousands of people and examined a range of topics, including the psychology of lying, how a jury is swayed by a defendant's appearance, and whether people can tell the difference between cheap and expensive wine (they couldn't).

A few years ago I arranged for thousands of people across Britain to take part in a large-scale study into happiness. Psychologists have created all sorts of techniques for promoting happiness, and I wanted to discover which was the most effective. Also, because other research had shown that happiness can spread through groups of people, like infectious disease, with people "catching" emotions from one another, I wondered whether thousands of happier people might act as a catalyst and cheer up the entire country.[17]

Before the start of the study I commissioned a national survey to measure the mood of the country. All of the respondents were asked to rate how cheerful they felt on a seven-point scale, where 1 corresponded to "not at all cheerful" and 7 to "very cheerful." Forty-five percent of the population awarded themselves a 5, 6, or 7.

The study was then announced in the national media. Everyone who was interested in participating was asked to visit the project's website and rate how happy they felt. More than 26,000 people responded. All of the participants were randomly assigned to one of a handful of groups and asked to carry out various exercises designed to make them happier. A number of the groups used some of the most popular "think yourself happy" exercises, involving, for example, creating a sense of gratitude or reliving happy memories, while participants in one of the other groups were asked to follow James's advice and smile for a few seconds each day.

A week later, the participants came back to the website and again rated how happy they were. When it came to increasing happiness, those altering their facial expressions came out on top of the class—powerful evidence that the As If principle can generate emotions outside the laboratory and that such feelings are long-lasting and powerful.

After the study, we conducted another national happiness poll. People were again asked to rate how cheerful they felt on the seven-point scale, and this time 52 percent put themselves in the top half. Assuming there are 60 million people in the country, this 7 percent rise corresponds to just over 4 million people reporting that they felt happier after the study. Was the increase due to our project? It is impossible to know for sure, but there were no obvious changes in the other factors that might have affected the mood of the country, such as a sudden rise in the amount of sunshine, drop in rainfall, or particularly heartening news stories. So we like to think that William James helped cheer up an entire nation.

III. The Value of Fun

William James not only speculated that smiling makes people feel happier, but also that all aspects of behavior, including the way people move and speak, would influence how they felt. To discover if he was correct, psychologists started to walk the walk and talk the talk.

Research shows that in the same way that there are only a very small number of core facial expressions, so there are only six basic walking styles. Striders, for example, take long steps, walk with a bounce, and let their arms swing back and forth. In contrast, shufflers tend to take small steps and have drooping shoulders. The work also showed that people associate each of the walking styles with different emotions, with striders being perceived as happy and shufflers as sad.

Psychologist Sara Snodgrass from Florida Atlantic University wanted to discover whether changing the way people walked would influence how they felt.[18] While pretending to be conducting a study on the effect of physical activity on heart rate, Snodgrass asked people to take a three-minute walk in one of two ways. Half of the participants were asked to take long strides, swing their arms, and hold their head up high. In contrast, the others were asked to take short strides, shuffle along, and watch their feet. After enacting this real-life version of Monty Python's Ministry of Funny Walks, everyone rated how happy they felt. The results demonstrate the power of the As If principle, with those who had been asked to take long strides feeling significantly happier than those who had been asked to shuffle.

The As If principle can also help bring people closer together moments after they meet. Sabine Koch from the University of Heidelberg is fascinated by the impact of movement on the mind, and her work into the psychology of dance has revealed that people feel happier when they move in a fluid way and unhappier when they make sharp and straight movements.[19] Aware that it isn't easy to persuade people to get in touch with their inner gazelle in everyday life, Koch turned her attention to a more down-to-earth behavior: handshaking.

Koch trained a group of experimenters to shake people's hands in one of two ways. Some of them learned how to shake hands in a smooth, flowing way, while others were shown how to produce sharper up-and-down movements. This crack team of intrepid handshakers then shook the hands of almost fifty participants. After each shake, Koch asked the participants how they felt. The results were remarkable: compared to those subjected to the spiky handshakes, those who had been subjected to the smooth, flowing shake were happier, felt psychologically closer to the experimenter, and rated the experimenter as more likable and open. The smooth handshake had made participants behave in a way that is associated with happiness, and this in turn had made them both feel better and think more of the person they had just met.

Other work has examined whether the words that you say and the way that you say them also influence how you feel. In the late 1960s American clinical psychologist Emmett Velten wanted to create a quick and easy way of generating good cheer in the laboratory.[20] What would happen, he wondered, if people spoke as if they were happy and confident? To find out, he assembled a group of volunteers, randomly split the participants into two groups, and handed each group a stack of cards.

For the first group, the top card in the stack explained that they were about to see a series of statements and were required to read each statement out loud. The next card contained the first of the statements: "Today is neither better nor worse than any other day." As instructed, the participants read the statement out loud and then turned over the card and moved on to the second statement: "I do

23

feel pretty good today, though." Slowly but surely, the participant moved through all sixty cards, with the statements becoming increasingly positive.

Those in the second group were asked to read a series of statements that were not designed to get them talking like a positive person and so spent the sessions reading various facts out loud—for example: "Saturn is sometimes in conjunction, beyond the sun from the earth, and isn't visible," "The *Orient Express* travels between Paris and Istanbul," and "The Hope diamond was shipped from South Africa to London through the regular mail service."

At the end of the procedure, Velten asked all of the participants to rate how happy they felt. Those who had said positive statements about themselves were in a wonderful mood. In contrast, those who had been reflecting on Saturn, the *Orient Express,* and the Hope diamond flat-lined.

Encouraged by Velten's results, other psychologists quickly adopted the procedure, and it is now used to cheer up experimental participants across the world.[21]

And it isn't just about reading single statements. In another study, Elaine Hatfield from the University of Hawaii and her colleagues had a group of participants read a short paragraph describing a fictitious scenario in which their friends had thrown them a wonderful surprise birthday party.[22] In contrast, another group read a paragraph describing how they had just heard that a member of their family had been diagnosed with an illness. Saying the two different sets of words affected the participants' moods, with the people hearing about a good time feeling much better than those speaking about the family illness. Getting the participants to speak as if they were in a good or bad mood had the power to genuinely affect their emotions.

The As If principle is not just about forcing your face into a smile. It applies to almost every aspect of your everyday behavior, including the way that you walk and the words that you say. Excited by these findings, academics quickly set about exploring other ways of using the As If principle to cheer people up in an instant.

Happy Talk

Can you really talk to yourself and make yourself feel happier? Find out by doing the following two exercises.

First, read each of the following statements out loud to yourself. Try to sound as convincing as possible, as if you are spontaneously saying the statements to a friend. Don't rush the task, but rather speak slowly and leave a little time before moving on to the next sentence. Most people find the task odd at first but quickly get used to it.

1. I feel surprisingly good about myself today.
2. I think that I can make a success of things.
3. I am glad that most people are very friendly toward me.
4. I know that if I set my mind to something, it will usually turn out well.
5. I feel very enthusiastic right now.
6. It's as if I am full of energy at the moment and enjoying what I am doing.
7. I feel especially efficient today.
8. I'm very optimistic at the minute and expect to get along very well with most of the people I meet.
9. I'm feeling very good about myself and the world today.
10. Given the mood that I am in, I feel particularly inventive and resourceful.
11. I am pretty certain that most of my friends will stick with me in the future.
12. I feel my life is very much under my control.
13. I am in a great mood and want somebody to play some wonderful music.
14. I am enjoying this, and I really do feel good about myself.

15. This feels like one of those days when I'm raring to go!

How do you feel now? For most people the procedure results in a boost in happiness.

Now try reading aloud the following paragraph. Again, try to make your words sound natural and enthusiastic. It might help to imagine that you are having a telephone conversation with a friend. Feel free to improvise and come up with your own positive version of events.

> This has been a great day. It's my birthday, and you will never guess what happened. I was invited to my friend's house earlier this evening, and when I arrived, I found out that he had arranged a surprise birthday party for me! It was amazing. Almost everyone I knew was there, and some people had made a real effort to be part of the event. They had baked me a cake, brought me presents, and even sang "Happy Birthday." I will remember the day for the rest of my life and am so lucky to have these friends.

Make 'Em Laugh

In 1995 Madan Kataria was working as a family physician in Mumbai, India. While researching a magazine article on the science of laughter, he learned about the medical benefits of laughter and decided to try to introduce more chuckles and guffaws into people's lives.

Kataria came up with a strange plan. At seven o'clock one morning, he went to his local park and persuaded four people to tell each other jokes and laugh. Everyone enjoyed the session, and Kataria decided to repeat the exercise the following week. The group soon grew, and more than fifty people were turning up for the sessions. Kataria had created the world's first laughter club.

In his original meetings, everyone would stand in a circle and then take turns telling a joke. Initially all went well, but after a few weeks, people ran out of clean jokes and started to use more questionable material. The laughter came to a sudden halt when two women threatened to leave because of sexist jokes, causing Kataria to explore other ways of putting a smile on people's faces.

Eventually he had a world-changing moment: he wondered whether people would get the same benefits from laughter if they chuckled away without hearing any jokes. The laughter club was initially skeptical but eventually agreed to put their mother-in-law jokes on the back burner and try Kataria's new approach. Within moments of acting as if they had heard a great joke, many members of the group found themselves feeling surprisingly euphoric. Contagion quickly kicked in, and soon almost everyone was giggling. Word about Kataria's new and surprisingly effective way of inducing good cheer quickly spread, and laughter clubs sprang up across the world.

Intrigued, psychologist Charles Schaefer from Fairleigh Dickinson University in New Jersey decided to discover whether behaving like someone who has just heard a great gag really does make people feel good.[23] He set up his own experimental laughter club and compared the effects of laughing with smiling.

Schaefer split volunteers into three groups. One group was asked to spend a minute smiling, and another was required to spend the same amount of time laughing out loud. Concerned that any changes reported by those in the second group may have been due to the physical exertion caused by side-splitting laughter, Schaefer wanted the third group to carry out a task that was equally as energetic but not associated with having a good time. After much head-scratching, Schaefer asked the students in the third group to howl like a wolf.

Although a clever control, the "howl like a wolf" exercise was not without its problems. At first the students in this group were somewhat confused and uncertain about how best to contact their inner

27

wolf. To solve the problem, Schaefer stood in front of the group and personally demonstrated how to role-play a wolf howling at the moon. He later reported that seeing a senior professor behaving in this way quickly caused the students to become significantly less self-conscious.

After much smiling, laughing, and howling, Schaefer asked everyone to assess their mood. The more the students behaved as if they were having a good time, the happier they became. Those who smiled became happier, and those who laughed out loud became elated. Howling like a wolf had very little effect on happiness, showing that the effect of the laughter was not due to physical exertion. William James's theory had been proven accurate again. (Schaefer didn't examine whether the students in the third group felt strangely attracted to dog food or nervous about silver bullets.)

Schaefer's study revealed why laughter clubs are so popular. In the same way that smiling makes you feel happy, behaving as if you find something funny brings the same psychological and physical benefits associated with genuine laughter.

Laugh and the World Laughs with You

Laughter clubs vary, but here is a rough guide to some of the basic procedures and exercises.

First, the group starts by forming a circle, with each person standing a few feet away from one another. One member of the group then takes the role of the leader and stands in the middle of the circle.

The entire session lasts around twenty minutes and is made up of various exercises, each of which lasts about forty seconds. Here are some of the more popular exercises.

The big "ho-ho ha-ha." Everyone starts chanting, "Ho-ho ha-ha," clapping their hands together on each "ho" and "ha." The chants should come from the stomach rather than the mouth, and everyone should try to keep a smile on their face throughout the exercise. This is often used as a warm-up and between other exercises.

Shake it all about. Everyone in the circle joins hands. The leader then says, "Now," and the group starts to quietly giggle. The leader beckons everyone toward the center of the circle. As they move forward, the intensity of their laughter increases. When the group nears the center of the circle, the leader motions them back, and they return to their original positions and their quiet giggling.

Taming the lion. Everyone adopts a "lion posture" by sticking their tongue fully out, opening their mouth and eyes as wide as possible, and holding their hands up in the way a lion holds its paws up. On a command from the leader, everyone roars like a lion for twenty seconds.

The hummingbird. Everyone pairs up, closes their lips, and tries to laugh while making a humming sound. Throughout the exercise, people try to maintain eye contact with their partner.

Mockery. The leader separates the circle into two groups. The groups look at one another and start laughing, often pointing at members of the other group. This exercise is not recommended if members of the group suffer from low self-esteem, paranoia, or both.

After conducting some surprisingly serious science into laughter, psychologists turned their attention to the effect of other equally enjoyable experiences, such as dancing the night away.

Happy people like to dance, but can dancing make you feel

happy? To find out, Sungwoon Kim from Kyungpook National University in Korea enlisted the help of approximately three hundred students.[24] The researchers split the students into four groups. Group 1 was asked to take part in an hour-long aerobic exercise class, the second group was invited to a body-conditioning session, group 3 had fun hip-hop dancing, and group 4 went ice skating. After taking part in the activities, everyone completed a questionnaire about their mood. It is well known that exercise makes people happier because it releases feel-good hormones called endorphins, and so the researchers expected all of the participants to be happier after the exercise. However, they wondered whether the hip-hop classes would make the participants feel especially good because they were behaving as if they were happy people. The results revealed that those who were sent hip-hop dancing were at the top of the class in terms of happiness.

And it isn't just hip-hop dancing that makes you feel good. Peter Lovatt, a colleague of mine at the University of Hertfordshire, carries out research into dance. Labeled "Dr. Dance" by the British media, Peter has examined a wide range of dance-related issues, including whether people with symmetrical bodies are better dancers (they are) and why fathers dancing at weddings make us cringe (they have an overinflated opinion of their abilities). A few years ago, Peter ran a ten-week experiment examining the effects of dance on mood.[25] Each week he gathered together a group of willing volunteers at the university, taught them a new type of dance, and then asked them to rate their mood. From the foxtrot to flamenco, and salsa to swing, everyone had a jolly time. Once again, the results revealed that behaving as if they were happy made the participants feel better, with dances that were noncompetitive and with an easily learned repetitive structure, such as Scottish country dances and line dancing, proving especially effective.

If you have two left feet, worry not: you can always sing a happy song. The seventeenth-century Spanish novelist and poet Miguel de Cervantes thought so, noting, "He who sings frightens away his ills." But was Cervantes correct?

Musicologist Grenville Hancox from Canterbury Christ Church University is a world-class clarinet player, conductor, and researcher. Curious about the effect that music has on people, Hancox has carried out several large-scale investigations into whether singing makes people happy, including one study in which he interviewed more than five hundred choral singers. The work has revealed one clear finding: singing makes people feel happy.[26]

Gunter Kreutz from the J. W. Goethe-Universität in Frankfurt had tackled the same issue under more rigorous circumstances.[27] Kreutz visited a choir during its rehearsal, asked the choristers to sing sections from Mozart's Requiem, and then had them rate how happy they were. As a control, Kreutz again gate-crashed their rehearsal a week later, asked them to listen to a recording of the same piece, and had them again rate their happiness. Although listening to the music didn't make people feel any happier, singing made them feel far more cheerful.

The message from research into the As If principle and the fun factory is clear: rather than trying to cheer yourself up by thinking happy thoughts, it is far quicker and more effective to simply behave as if you are having a good time. Smile, put a spring in your step, hold your head up high, use happy talk, dance, laugh, sing, or do whatever else you enjoy doing.

Or, to put it another way, if you want to be happy and you know it, clap your hands.

Chapter 2
Attraction and Relationships

Where we unpack the mysteries of the human heart, discover the power of footsie, invent a new type of speed dating, and learn how to live happily ever after

"Whatever we learn to do, we learn by actually doing it; men come to be builders, for instance, by building, and harp players by playing the harp. In the same way, by doing just acts we come to be just; by doing self-controlled acts, we come to be self-controlled; and by doing brave acts, we become brave."

—Aristotle

I. What Is Love?
II. The Misattribution of Bodily Sensations
III. Love in the Laboratory

I. What Is Love?

In 1981 Prince Charles announced his engagement to Diana Spencer. In a now infamous television interview about their forthcoming wedding, journalist Anthony Carthew asked the couple how they felt. Charles hesitantly explained that he was delighted and happy, prompting Carthew to add, "and I suppose . . . in love?" Diana quickly agreed, but Charles was far more circumspect, mumbling, "whatever 'in love' means."

Charles is not the first to be confused by the nature of love. Throughout history, poets, musicians, and writers have struggled to define this most elusive of emotions. The ancient Greek philosopher Aristotle thought that love was best viewed as "a single soul inhabiting two bodies," whereas Elizabeth Barrett Browning attempted to capture the essence of passion when she wrote: "What I do and what I dream include thee, as the wine must taste of its own grapes." In contrast, the American actor John Barrymore took a somewhat more practical view, noting that "love is the delightful interval between meeting a beautiful girl and discovering that she looks like a haddock."

Although it is difficult to define love, there is little doubt that the emotion has always intrigued us. Archaeologists working in Iraq's Niffer Valley recently unearthed the world's oldest surviving love letter. Etched into a four-thousand-year-old clay tablet, the love poem appears to have been written by a high priestess to the man she is about to marry and describes her excitement about their forthcoming wedding night ("Goodly is your beauty, honeysweet. You have captivated me. Let me stand tremblingly before you. I am even prepared to buy you a reasonably priced slice of cheesecake.")*.

Love also knows no cultural bounds. From the Amazon to Arizona, and the Sahara to Siberia, people in vastly different societies all appear to experience the joys and pains of passion. The few communities that have attempted to banish love have encountered nothing but failure. In the nineteenth century, for instance, both the Shakers and the Mormons believed that love was lust in disguise, and so attempted to prevent their followers having romantic encounters. Love persisted in both groups, often taking the form of dangerous liaisons away from prying eyes.

Given the ubiquity of love, one might have expected psychologists to have had a long-standing interest in the emotion. However, research into the mysteries of the human heart is a surprisingly recent endeavor and was kick-started by the strangest of events.

"I'm Just One of You in a Bag"

In 1967 Professor Charles Goetzinger was lecturing on the science of persuasion at Oregon State University (the course was Speech 113: Basic Persuasion). When his new cohort of students arrived for the first of their lectures, they were greeted by a strange sight.[1] Sitting at one of the desks was a person completely covered in a large black cloth bag, with bare feet poking out through two slits in the base of the bag.

Goetzinger explained to his class that a male student had decided to attend the course enclosed within a black bag and wished his identity to remain a complete secret. In the absence of a name for their anonymous classmate, the students agreed to refer to him as "Black Bag" (presumably having just skipped Speech 112: Very Basic Creativity).

The class met three times a week, and each time Black Bag sat silently at his desk. When the time came for the students to give a

---

* I added the final line.

three-minute presentation on persuasion, Black Bag simply stood in front of the class in total silence. Initially Goetzinger's students were antagonistic toward Black Bag. One poked him with an umbrella, another placed a "kick me" label on his back, and a third tried to punch him.

The story soon caught the attention of the local, and then national, media. News crews from across America descended on Goetzinger's classes; TV anchor Walter Cronkite tried to interview the mysterious man in the bag, and *Life* magazine devoted several pages to the story.

Then something unexpected began to happen. After a few weeks, the students started to bond with Black Bag. Despite still not knowing who he was or what he looked like, the punching and poking were replaced with compassion and kindness. As alienation evolved into acceptance, the students grew increasingly fond of their anonymous classmate, involving him in their activities and protecting his identity. When Goetzinger asked the class to vote on whether Black Bag should be asked to reveal his identity, the majority of students were opposed to the idea.

At the end of Goetzinger's persuasion course, several camera crews were lined up outside the school building waiting for Black Bag to emerge from class. Without a word said, the students formed a human wall around Black Bag and protected him as he moved through the media scrum. The episode humbled Black Bag, causing him to utter a rare but memorable comment: "I'm just one of you in a bag." To this day, the identity of Black Bag remains a mystery.

Both the media and public turned to psychologists in an attempt to understand why Goetzinger's students had slowly warmed to their anonymous classmate. There was, however, one small problem: psychologists at the time had almost no idea how to explain what had happened.

Prior to the 1960s, most psychologists saw the experimental examination of friendship, attraction, and love as taboo. Perhaps eager to distance themselves from Freud's unscientific and overly sexual view of the human psyche, universities discouraged their staff from investigating people's personal lives. Entering the forbidden zone could have very real consequences. In fact, one professor was severely reprimanded for carrying out a survey in which he asked people whether they had ever blown into someone's ear in an attempt to arouse passion.[2]

Even by the start of the 1960s, researchers had developed only the most rudimentary of ideas about how people come to like and love one another. Somewhat embarrassed by the lack of insight that their work could shed on the mystery of Black Bag, a handful of researchers began to wander into the academic wilderness and investigate the psychology of friendship, attraction, and love.

In 1975, psychologist Elaine Hatfield from the University of Wisconsin received funding from the National Science Foundation to carry out one of the first systematic programs of research into love and attraction. Although many researchers saw this as a major breakthrough, not everyone was thrilled. Senator William Proxmire launched a blistering attack on Hatfield's work, awarded her one of his famous Golden Fleece Awards (for allegedly fleecing taxpayers), and publicly announced:

> I believe that 200 million other Americans want to leave some things in life a mystery, and right on top of the things we don't want to know is why a man falls in love with a woman and vice versa. . . . So National Science Foundation—get out of the love racket. Leave that to Elizabeth Barrett Browning and Irving Berlin. Here if anywhere Alexander Pope was right when he observed, "If ignorance is bliss, tis folly to be wise."

Unperturbed, Hatfield plowed on. In one of her initial studies[3] she teamed up with fellow romance researcher Russell Clark from the Florida State University and examined a very straightforward question: If a reasonably attractive member of the opposite sex asked a man or woman to sleep with him or her, would the person get lucky?

Hatfield and Clark asked five women and four men to walk up to complete strangers at the university and say: "I have been noticing you around campus. I find you very attractive. Would you go to bed with me tonight?" The experimenters carefully recorded the response in a notebook and then explained that they were actually conducting a social psychology study and thus their offer had been made solely in the spirit of scientific inquiry (the experimenters did not record people's responses to this part of the procedure). Describing their findings in a paper, "Gender Differences in Receptivity to Sexual Offers," Clark and Hatfield reported major differences between the sexes. None of the women who were approached accepted the offers of sex from the male experimenters. In contrast, a remarkable 75 percent of men checked the "your place or mine?" box.

It's perhaps not surprising that Hatfield's study generated a storm of controversy, with some arguing that the findings provided a vivid demonstration of how those with societal power exploit the powerless and others being equally adamant that they provided support for the "man = shallow" hypothesis. The study has also had an unintended effect on popular culture. In 1998 the British jazz ensemble Touch and Go used Hatfield's experimental script as the basis for their song "Would You . . . ?" The song reached number three on the U.K. singles chart and has been viewed 2 million times on YouTube.

The success of this initial work prompted Hatfield and her colleagues to conduct other experiments into the psychology of attraction. Some of the subsequent work revealed that both friendship and love tended to be greatly enhanced by prolonged contact. According to this theory, the more you hang around someone, the greater the likelihood of that person liking and eventually loving you. This principle was used to explain why people often end up marrying people from their neighborhoods and why Goetzinger's students slowly came to bond with Black Bag. The theory also apparently motivated one man to write more than seven hundred letters to his girlfriend, resulting in her marrying the postman (only joking).

The trickle of research into love soon became a tide, and since the mid-1970s, hundreds of researchers have conducted thousands of experiments in an attempt to unravel the mysteries of the human heart.[4]

Love Factually

Research into Cupid's deepest secrets took many forms, including secretly observing people flirting in singles bars, staging scientific speed-dating sessions, posting fake personal ads, testing testosterone levels during kissing, and examining the lives of happily married couples.

It soon became apparent that love was going to be a tough nut to crack. In the early 1970s, for example, psychologist Donn Byrne declared that he had discovered the equation for love, proudly announcing that

$$Y = m[\sum PR/(\sum PR + \sum NR)] + k,$$

where Y stands for attraction, PR for positive reinforcement, NR for negative reinforcement, and k is a constant.[5] The response to Byrne's alleged discovery can be summed up by another formula,

$$X < 1,$$

where X stands for the number of people who were impressed.

Other academics took a slightly more constructive approach, arguing that we all unconsciously carry around a list of desirable traits in our head, and when we meet someone who checks all of the boxes, our brain suddenly goes into overdrive and we fall in love.

Although the work struggled to discover exactly how Cupid's arrow found its target, it did reveal that there are two main varieties of love.[6] The first, referred to as "passionate love," is associated with feelings of intense exhilaration, infatuation, and an emotional high. It is the type of love that makes two people stay up all night chatting and watch the sun rise in the morning. Some psychologists have taken a romantic view of this experience, focusing on the upside of a couple yearning to be together and thinking endlessly about each other. Others have taken off their rose-tinted glasses and adopted a more down-to-earth approach, pointing out that passionate love activates the parts of the brain that are commonly associated with drug abuse and alcohol addiction.

The second type of love, known as "compassionate love," is far more about attachment than attraction. Rather than focusing on the thrill of the chase and intensity of that first kiss, this form of love is experienced by those in a secure and comforting long-term relationship.

Over the years, researchers have devised questionnaires to measure these two very different varieties of love (see the "Are You in Love?" box). A few years ago, Elaine Hatfield and her colleagues gave the questionnaires to three types of couples: those who had just started dating, newlyweds, and long-term married.[7] The results allowed Hatfield to track how love evolves over the course of a relationship.

First, the good news: those who have just started dating experience a very high level of passionate love and a moderately high level of compassionate love. Next, the even better news: newlyweds report even higher levels of both passionate and compassionate love. Now the not-so-good news: within a year of being married, the rot starts to set in, with both passionate and compassionate love declining to the levels experienced during dating. Finally, the bad news: over the next thirty years of marriage, both passionate and compassionate love decline, with the demise of the former being far more rapid than the latter. Love may never die, but it certainly takes a dramatic nosedive as time passes. On the upside, as we will discover later in this chapter, once you understand the truth about love, it is relatively easy to maintain a sense of passion in even the longest of relationships.

Couple Hatfield's somewhat depressing findings with the human need to be loved, and it isn't surprising that throughout history, those offering ways of manufacturing long-lasting love have never been short of clients.

Are You in Love?

Please complete the following questionnaire by imagining your partner's name in each of the following statements and then assigning a number between 1 (definitely not true) to 5 definitely true). Jot down your answers on a piece of paper.[8]

1. I would feel heartbroken if _____ left me.
2. _____ is always on my mind.
3. Of all of the people I know, I would rather be with _____.
4. If _____ were to fall in love with someone else, I would feel extremely jealous.
5. I tingle all over when _____ touches me.
6. I feel very sad when I see _____ going through a difficult time.
7. _____ and I make a great team.
8. Helping _____ provides me with a strong sense of meaning to my life.
9. I would rather help _____ than help myself.
10. I feel very comfortable when I am together with _____.

Scoring

Add up your answers to questions 1 to 5. This is your passionate love score. Use the following chart to discover how your score compares to others.

Less than 5	You didn't read the instructions properly.
5–7	Extremely cool; the thrill seems to be gone.
8–10	Feels somewhat cool and tepid.
11–15	Average, with occasional periods of passion.

| 16–20 | Passionate but still room for improvement. |
| 21–25 | Congratulations! You are wildly and passionately in love. |

Next, add up your answers to questions 6–10. This is your compassionate love score. Use the following chart to find out how your score compares to others:

5–7	Slight evidence of friendship.
8–10	The two of you have bonded, but it isn't strong.
11–15	Average, with the occasional burst of strong compassion.
16–20	Highly compassionate.
21–25	Congratulations! You are in compassionate love.

In Search of Cupid

For centuries, magicians and witches claimed to be able to create spells and potions that could make people fall in love. The Byzantines believed that Cupid's arrow would strike the moment two people consumed a special "love cake" made from donkey's milk and honey. Throughout the Middle Ages, the humble tomato was widely seen as the "apple of love," causing Puritan leaders to try to prevent their followers from eating the vegetable by spreading the rumor that it was poisonous. When spells, donkey's milk, and tomatoes failed to hit the mark, people turned to more down-to-earth ways of finding the love of their life, including personal ads.

Personal ads have a surprisingly long history, with the first one appearing in 1695 in a British publication entitled *The Collection for the Improvement of Husbandry and Trade*.[9] The item was sandwiched between an advertisement for an Arabian stallion and a second-hand bed and was placed by a well-heeled gentleman who was searching for a "Gentlewoman that has a fortune of £3,000 or thereabouts." History doesn't record whether the young man's quest was successful. Nevertheless, the idea quickly caught on, with people placing increasingly demanding and humorous ads. One eighteenth-century ad appealed for respondents with "no bodily deformity," another wanted only those with a "shapely ankle," and a third advertised for "a wife with two or three hundred pounds; or the money will do without the wife."

Skip forward a few hundred years and the search for love shows no signs of fading away. In 1965, three enterprising Harvard students created Operation Match, the first computer-based matchmaking system. To test the system, the students asked more than seven thousand people to complete questionnaires designed to assess their personalities. This information was then transferred onto punch cards and run through a mainframe computer that was the size of a small room. Six weeks later, each participant received a list containing the addresses and telephone numbers of dates thought to be compatible. Nowadays, online dating sites employ the same concept and have developed increasingly sophisticated algorithms to match accurately millions of people using a variety of factors.

The matchmaking industry's latest innovation was created in the late 1990s when Rabbi Yaacov Deyo wanted to help Jewish singles meet one another and so came up with the concept of speed dating. The idea quickly caught on, leading to millions of people around the world trying to find love on the basis of a three-minute chat.

The love industry does not just cater to those in search of Cupid's arrow. If couples experience difficulties in their relationship, counselors, DVDs and self-help books offer a wide range of advice that claims to help people stay together.

But are these techniques effective? Research suggests that there is about a 4 percent chance of having a relationship with someone you have met at a speed-dating event.[10] Online dating fares better, with one survey carried out by a major online dating agency showing that about 17 percent of couples who tied the knot in the three years prior to the survey met through an online dating site.

The recent figures for those wanting to stay in love are somewhat more depressing, with around half of Americans' first marriages failing, along with two-thirds of second marriages and three-quarters of third marriages.

Encouraged by work showing how the As If principle can be used to manufacture happiness, researchers wondered whether it could also help people find the love of their life and encourage couples to stick together when the going gets tough.

Cue Dr. Stanley Schachter and his hula hoop of love.

II. The Misattribution of Bodily Sensations

Think about the last time you experienced a really strong emotion. Perhaps you felt anxious before giving a talk, nervous during an important job interview, excited after a successful date, or angry when someone insulted you. In any case, unless you are a psychopath, you probably noticed a dramatic change in your bodily sensations. Your heart rate may have increased, your mouth might have become extremely dry, and perhaps your palms felt sweaty.

Much of the initial work into the psychology of emotion attempted to identify the exact patterning of bodily sensations that accompany various feelings. Researchers invited participants into the laboratory, wired up their bodies to various sensors, and then made the participants angry by insulting them, scared by bombarding them with loud noises, and happy by offering them cake. The research teams then pored over the reams of data, looking for the pattern of bodily sensations associated with each emotion. Was anger associated with a sudden increase in heart rate coupled with faster breathing? Was fear linked with a very dry mouth and sweating? Was joy accompanied by a decrease in heart rate and much shallower breathing?

After spending years trying to create a physiological dictionary of emotion, it was obvious that something was very wrong: although almost all of the participants experienced a vast array of emotions, the bodily sensations that accompanied these feelings were often surprisingly similar. Something just didn't add up.

Then, in the 1960s, psychologist Stanley Schachter finally solved the mystery. Schachter worked at Columbia University and studied a wide variety of interesting topics, including obesity, nicotine addiction, cults, and miserliness. Early on in his career, he conducted a now-classic experiment exploring what happens inside your body when you experience an emotion.[11] Let's imagine that you are taking part in his study.

You're walking along the street minding your own business, when suddenly you see a poster asking for participants to take part in an experiment investigating the effects on vision of the vitamin compound suproxin. Eager to earn some money for a few hours' work, you call the number on the poster and are asked to come along to Schachter's laboratory the following day.

When you arrive at the laboratory, a researcher gives you an injection of suproxin, explains that the shot takes time before it has any effect, and asks you to go to a nearby waiting room. When you enter, you smile politely at a man already sitting there. The two of you start chatting, and the man explains that he is also a participant in the experiment. Like you, he is waiting for the injection of suproxin to take effect.

After a few minutes, your newly found friend becomes full of the joys of spring. He finds a hula hoop in one corner of the waiting room and starts playing with it, cracks several jokes, climbs on the furniture, and throws balls of paper into the wastepaper basket. After you have spent about fifteen minutes with Mr. Euphoric, the researcher enters the waiting room and asks you to complete a short questionnaire about your current mood. When you have completed the form, the researcher explains that the experiment has finished. But as is often the case with psychology, nothing was quite as it seemed.

Schachter was convinced that the scientific search for the bodily sensations associated with emotions had failed because it was based on a fundamentally flawed assumption. To Schachter, it seemed obvious that each emotion couldn't possibly be associated with a particular pattern of heart rate, breathing, sweating, and so on. There were simply too many emotions and too few bodily sensations. Instead, Schachter thought that the situation was far simpler. He hypothesized that all bodily sensations are caused by a physiological system that operates much like a tug-of-war match.

At one end of the rope is the red team. When this team bursts into action, you feel aroused and active. Adrenaline and sugar are quickly released into your bloodstream to help provide energy, your heartbeat and breathing rate increase to get more oxygen to your muscles, there is a reduction in blood flow to your skin to help reduce bleeding if you are injured, and the digestive juices in your stomach go into overdrive to help produce more energy. In short, your body undergoes the well-known fight-or-flight response. If you decide not to engage in fisticuffs or run away, the unspent energy in your body can make you feel light-headed and weak at the knees; you experience butterflies in your stomach, and you tremble.

On the other end of the rope is the blue team. When they pull on their end of the rope, your body calms down. Your heart rate slows, and your digestive system returns to normal. When you lie down and take it easy, the blue team pulls on their end of the rope, causing your heart rate to decrease and your breathing to become slow and shallow. The moment you stand up and walk about, the red team bursts into action and restores your heart rate and breathing to normal.

Most of the time, the red and blue teams work together to ensure that your bodily sensations are appropriate for your surroundings. If, for example, you were to spot a tiger in some bushes, the red team would jump into action and you would feel your heartbeat suddenly increase. However, the moment you remember that you are at a zoo and therefore perfectly safe, the blue team would pull on their end of the rope, and your heartbeat would slow back down.

According to Schachter, there is not a different bodily reaction associated with each emotion but rather one system that simply varies in intensity.

Body Talk

People vary in the degree to which their bodies produce the types of sensations that drive emotional feelings. Use this questionnaire to discover the degree to which your body is responsive.[12]

Imagine that you are in a fairly stressful situation. Please rate the degree to which each of the ten statements would apply to you using the following scale:

1: never; 2: occasionally; 3: sometimes; 4: usually; 5: always.

During stressful situations . . .

... my face either blushes or becomes especially pale.

... my legs feel weak and my hands start to shake.

... my breathing becomes rapid and shallow.

... my heartbeat increases.

... my palms start to sweat.

... my stomach starts to rumble.

... the hair on the back of my neck stands up.

... my mouth feels dry.

... my eyes start to feel wet.

... my face and ears feel hotter.

Scoring

Add up your scores. Use the following chart to find out how your score compares to that of others.

10–20   Very low reactivity
21–30   Above low but below average reactivity
31–40   Average reactivity
41–50   Very high reactivity

> Being a high or low reactive is neither inherently good nor bad. Low reactives tend to keep calm in stressful situations, whereas high reactives are good at responding to early signs of threat.

Schachter's theory faces one major problem: If your bodily sensations vary only in intensity, how can you experience such a vast array of emotions? His solution involves moving away from your body and into your brain. According to his theory, when you experience a surge in bodily activity, you look around, try to work out what is going on, and then label the emotion accordingly. So if, for example, someone shouts at you, you would feel your heart beat faster, hear some insults, and conclude that you must be angry. Similarly, when you are with someone you find very attractive, you might experience exactly the same increase in heart rate but instead assume that it is due to desire.

Schachter's idea turned the commonsense view of emotion upside down. According to common sense, emotion seems to precede bodily sensations. You see a lion, become afraid, and start to sweat. Or you see a roller-coaster, become excited, and your heart beats faster. Schachter suggested the exact opposite: the lion makes you sweat; then you look at the dangerous situation you are in and start to feel afraid. Or the sight of the roller-coaster makes your heart beat faster; then you see that you are in a theme park and so feel excited. Schachter's idea is an extension of William James's theory of emotion. James believed you monitor your facial expressions and behavior and then figure out how you are feeling. Schachter extended this idea to your bodily sensations (see diagram).

Bodily sensations and emotion

Common sense suggests that the chain of causation is:

See an oncoming car— Feel afraid— Stomach lurches

Walk past your favorite celebrity in the street— Feel excited— Start to sweat

Schachter's theory suggests that the reality is:

See an oncoming car— Stomach lurches— Look at your situation— Feel afraid

Walk past your favorite celebrity in the street— Start to sweat— Look at your situation— Feel excited

---

This extension of the As If principle, if correct, leads to an intriguing prediction: it should be possible to get you to experience vastly different emotions by increasing your heart rate in different contexts. And this was exactly what your time with Mr. Euphoric was all about.

The vitamin compound "suproxin" doesn't exist and the study was nothing to do with vision. Also, as you may have figured out by now, Mr. Euphoric was actually a stooge working for the experimenters.

The injection that you received during the experiment contained a shot of adrenaline so would have activated your physiology. Moments after the chemical entered your body, the red team would have burst into action, making your heart beat faster, your hands tremble, and your mouth become dry. According to Schachter's version of the As If principle, spending time with Mr. Euphoric would have encouraged you to label your strange bodily sensations as signs of good cheer, and so made you feel especially happy.

And this is what the results revealed. Time and again, when participants who had hung out with Mr. Euphoric completed Schachter's mood questionnaire, they checked the "I feel strangely delighted" box.

The As If principle suggests that if exactly the same bodily sensations were experienced in a different context, they would give rise to a very different emotion. To find out if this was the case, Schachter carried out the second part of his study. He invited a group of participants into his laboratory and injected them with "suproxin" (again adrenaline). The participants were then given a questionnaire about their background and asked to complete it in the waiting room.

When each participant entered the room, they didn't encounter Mr. Euphoric and his hula hoop of joy. Instead, the same man had been asked to play Mr. Angry—a very unhappy individual who started to complain about the questionnaire. Faced with a series of increasingly personal questions—"With how many men (other than your father) has your mother had extramarital relationships? Options: 4 and under ____, Between 5 and 9 _____, 10 and over _____"—Mr. Angry became progressively more agitated and eventually wadded up the form and stormed out of the room.

Would Mr. Angry's behavior cause the participants to attribute their high heart rate to annoyance? Yes. In fact, when this second group of participants described their mood, they felt furious.

In both parts of the study, people's bodily sensations were identical. However, in the first part, Mr. Euphoric encouraged people to view their high heart rate in a positive way and so ended up feeling happy. In the second part of the study, Mr. Angry put a damper on the situation, causing people to view their heart rate in a more negative way and end up feeling annoyed.

To ensure that participants' emotions were not just the result of playing with hula hoops or being asked about their mother's morality, Schachter added other parts to the experiment, including two more groups that were injected with an inert saline solution rather than adrenaline. These participants didn't feel their heart race and so were not acting as if they were especially emotional. As a result they didn't feel especially happy or furious after spending time with Mr. Euphoric or Mr. Angry.

The As If principle also explains many curious aspects of emotion, such as people crying when they are either extremely sad or deliriously happy. Conventional theories of emotion struggle to explain why such different emotions should produce exactly the same behavior. According to Schachter's approach, the bodily sensations associated with every emotion are the same, and so emotions of equal intensity provoke the same bodily response.

Curiously, several studies have also suggested that when the temperature rises, there is an increase in serious and deadly assaults.[13] Once again, the As If principle has no problem explaining this phenomenon. When people are in environments that they find unusually hot, their heart beats faster and they start sweating. Some people will look around for an explanation for this change in their physiology, misinterpret these signals as signs of their being angry, and behave accordingly. It is an interesting explanation, but is it correct? To find out, one group of researchers heated up their laboratory to the mid-nineties Fahrenheit and gave participants an opportunity to administer electric shocks to one another. The result was that the hotter the room, the higher the shocks were. The researchers then cooled people down by asking them to drink a bottle of cold water and gave them another opportunity to shock others. Suddenly they became far less aggressive.[14]

However, by far the most significant impact of the As If principle was on those wishing to make love, not war.

The Chemistry of Love

Schachter likened the relationship between bodily sensations and emotion to a jukebox. In the same way that a coin is necessary for the jukebox to operate, so an event makes the body jump into action. And just as the user selects which tune to listen to, so people unconsciously look at what is happening around them, decide how their bodily sensations should be interpreted, and experience an appropriate emotion. Following in the footsteps of James, Schachter had used his version of the As If principle to create happiness and anger. But could it be used to produce passion?

To find out, Gregory White from the University of Maryland and his colleagues carried out two groundbreaking studies into heart rate and love.[15] In both studies, White made men's hearts beat faster, showed them a videotape of an attractive woman talking about her hobbies, and then asked the men to rate how sexy she was and how much they would like to kiss her.

In his first experiment White had one group of men run on the spot for two minutes (high heart rate), while another group carried out the same exercise for just a few seconds (low heart rate). In the second study, the groups of men listened to a tape of comedian Steve Martin's "A Wild and Crazy Guy" routine or to a grisly account of a mob killing a foreign missionary (high heart rate) or a dull description of the frog's circulatory system (low heart rate).

As predicted, the men who had their heart rate increased by the two-minute run, Steve Martin, or the tale of the murdered missionary found the woman in the video far more attractive than those who had exercised for only a few seconds or heard about frog physiology.

There have been other confirmations too.[16] In perhaps the best known of these, psychologists Donald Dutton and Arthur Aron famously arranged for a female "market researcher" (actually a stooge) to approach men on one of two very different bridges across the Capilano River in British Columbia.[17] One bridge was swaying precariously in the wind while the other was far more solid. The precarious footbridge caused the men's heart rate to increase, and they misattributed this as a sign of passion and found the woman particularly attractive. In another study, psychologists Cindy Meston and Penny Frohlich from the University of Texas visited theme parks and interviewed people a few moments before or after they had been on a scary roller-coaster ride.[18] The researchers presented people with a photograph of a woman and asked them to rate how attractive they found her. Those rating the photographs after the ride misattributed their sweaty palms as a sign of love and found the woman especially attractive.

As I noted in my previous book, *59 Seconds*, this work has important implications for anyone who wants to find the love of her life: if you are going on a date, keep away from country walks and meditation classes, and instead head to theme parks, high bridges, comedy shows, and movie theaters showing scary movies about mutilated missionaries.

This aspect of the As If principle also helps to explain several other curious aspects of love.

O Romeo, Romeo, Wherefore Art Thou Romeo?

Unrequited love often increases a person's sense of longing for an unobtainable partner. Such effects can be dramatic, with one spurned lover eventually kidnapping his former sweetheart and later tearfully explaining, "The fact that she rejected me only made me want to love her more." Most psychological theories would struggle to explain these events because humans tend to go out of their way to avoid people who make them feel bad. However, the As If principle presents a possible explanation for the phenomenon.

When someone is prevented from doing something that he or she wants to do, that person tends to become frustrated and angry. If that person happens to be in love, she may misinterpret the physiological signs of her frustration as evidence of passion. The more this person is spurned, the more she is attracted to the unobtainable apple of her eye.

The theory also explains the rather strange effect that barriers to love can have on the human heart. The Greek writer Vassilis Vassilikos once created a story about two mythical creatures. One of the creatures was a fish that was a bird from the waist up, while the second was a bird that was a fish from the waist up. The two creatures were passionately in love with each other, and one day the Fish-Bird expressed its annoyance that the two of them could never live together (sex wouldn't have been easy either). Looking on the bright side of a tricky situation, the Bird-Fish replied, "No, what luck for both of us. This way we'll always be in love because we will always be separated."

Vassilikos is not the first writer to suggest that separation fans the flames of love. In the Roman myth of Pyramus and Thisbe, a couple falls in love, but their parents disapprove of the relationship and so try to prevent the lovers from seeing each other. Confined to separate rooms in adjoining houses, the lovers whisper to each other through a crack in the wall, with writer Edith Hamilton noting in her version of the story, "Love, however, cannot be forbidden. The more the flame is covered up, the hotter it gets." Similarly, in Shakespeare's famous tragedy *Romeo and Juliet,* the more the lovers' feuding families try to push Romeo and Juliet apart, the deeper their passion for each other grows.

To discover if this rather curious phenomenon exists in real life, psychologist Richard Driscoll from the University of Colorado tracked the lives of more than a hundred couples during the course of a year, measuring both their level of mutual love and any attempts by their parents to prevent the relationship.[19] The more the parents attempted to prevent a relationship, the greater the couple's love for each other developed. In honor of the great Bard, Driscoll labeled the phenomenon "the Romeo and Juliet" effect.

Most conventional theories of love would adopt an out-of-sight, out-of-mind approach to the issue and predict that separating two lovers would cause them to slowly lose interest in each other. In contrast, the As If principle has no problem explaining the phenomenon. The more lovers are kept apart, the angrier they feel, and so the more likely they are to misinterpret these feelings of frustration as signs of passion.

The As If principle may also explain the infamous rebound effect. When a relationship ends, people often feel especially anxious. If they meet a new potential partner soon after a previous relationship has finished, they may misinterpret their anxiety as a sign of passion. Evidence for this effect comes from a study in which researchers arranged for a group of men to take a personality test and receive either positive feedback (to make them feel good) or negative feedback (to make them feel anxious). The men were then asked to wait in a cafeteria; while they waited, an attractive woman approached them. The men who had just received the negative feedback found the woman especially attractive, as Schachter predicted.[20]

One further effect, the Stockholm syndrome, also adheres to Schachter's version of the As If principle. When people are unfortunate enough to be taken hostage, they often develop a strange sense of affection toward their captors. The effect is surprisingly common, with the FBI's Hostage Barricade Database System suggesting that just under a third of hostages show some evidence of the syndrome. Interestingly, the effect usually emerges only when the captors have shown some degree of kindness to the hostages and so may well be the result of hostages' misattributing the anxiety caused by being denied their freedom as a sign of liking. The same idea may also help to explain why some individuals are attracted to partners who treat them badly.

For years, psychologists believed that people's emotions affected their physiology—that feeling angry made their heart beat faster and that feeling anxious made them sweat. In the same way that research into James's theory proved that people's behavior causes them to experience emotion, so Schachter showed that the way in which people interpret their bodily sensations determines which emotion they feel. Depending on the context, the same thumping heart can be seen as a sign of anger, happiness, or love. Schachter's theory was used to create love by showing people funny movies, having them walk across precarious bridges, and ride on scary roller-coasters. The theory also explains many curious aspects of love, including why rejection leads to increased attraction, why trying to keep lovers apart fuels the flames of passion, and why some people find it difficult to walk away from partners who are abusing them.

Encouraged by this work, researchers began to explore other ways in which the As If principle influences Cupid's arrow.

III. Love in the Laboratory

Around the turn of the last century, the maverick Victorian scientist Sir Francis Galton devoted much of his life to the study of strange psychological phenomena.[21] Adhering to his mantra, "Whenever you can, count," he determined whether his colleagues' lectures were boring by measuring the level of fidgeting in their audiences, tested the power of prayer by calculating the average life span of priests, and spent months trying to brew the perfect cup of tea.

In an essay entitled "Measurement of Character," Galton suggested that it might be possible to calculate the degree to which two people are in love by recording the "inclination" that they have for each other.[22] Galton had noted that when two lovers sit at a dinner table, they visibly slope toward each other and in doing so place greater weight on the front legs of their chairs. The great scientist suggested that it might be possible to secretly incorporate a pressure gauge with an index and dial into the legs of everyday furniture and thus objectively measure the extent of love. Galton ends his discussion of the matter by noting, "I made some rude experiments, but being busy with other matters, have not carried them on as I had hoped." Victorian scientists were reluctant to modify their furniture along the lines Galton suggested, and the idea of measuring how lovers behaved faded.

In fact, the idea of trying to record the behavior of those in love didn't reemerge until the 1970s. Rather than adopt Galton's idea of concealed gauges and dials, the work took a more observational approach. For several years, a small group of dedicated researchers boldly went where few psychologists had gone before and visited bars and parties. Once there, the researchers secretly observed the behavior of couples in love. The results confirmed what most people would have suspected: those struck by the love bug move physically close to each other, spend a long time looking into each other's eyes, play footsie under the table, mimic each other's body language, touch each other's hands and arms, and share secrets.[23]

Inspired by the way in which the As If principle had been used to manufacture happiness, researchers wondered whether people would fall for one another if they behaved as if they were in love. One of the first studies was conducted by Kenneth Gergen from Swarthmore College.[24] Couples often spend quality time together in the dark, and so Gergen wondered what would happen if he encouraged complete strangers to do the same. Gergen first covered the floor and walls of a ten-foot-square room with padding and asked groups of four men and four women to spend an hour in the room. After that, Gergen turned out the lights in the room and asked the groups of participants to spend sixty minutes in total darkness.

Gergen used normal and infrared cameras to record what happened in the room and also interviewed each participant after the experiment had finished. Describing his findings in an article entitled "Deviance in the Dark," Gergen noted that when the lights were on, none of the participants purposefully touched or hugged one another and that 30 percent of them felt sexually aroused. When the group was plunged into darkness, the situation was very different. Now, almost 90 percent of them touched one another on purpose, 50 percent hugged, and 80 percent were sexually aroused. In addition, the people who were in the dark room were far more likely to start talking about important events in their lives and find one another attractive. Gergen's footage revealed that a few of the participants even started stroking one another's faces and kissing. Simply by finding themselves in the sort of situation that lovers enjoy, people quickly started to behave as if they had been struck by Cupid's arrow and so found one another more attractive.

When it came to making love in the laboratory, this was just the tip of the iceberg.

The Power of Footsie

Harvard psychologist Daniel Wegner wondered whether, in the same way that smiling makes people feel happy, having two strangers secretly play footsie might make them feel attracted toward each other.[25] Mindful of the way in which those investigating the "smiling makes you happy" hypothesis had used various cover stories to prevent participants telling the experimenter what they wanted to hear, Wegner pretended he was conducting a study examining the psychology of poker.

Volunteers were invited to Wegner's laboratory in groups of four. The experimenters scheduling the volunteers ensured that none of the group knew one another and that each foursome consisted of two men and two women. The researchers then split the volunteers into two teams, with each consisting of one man and one woman. The teams were told that they were going to play poker against each other, and the experimenters took the teams to two separate rooms to explain the rules of the game. In reality, one of the teams was taught how to cheat by secretly sending codes to each other during the game. And how were these codes relayed? The cheating couple were told to keep their feet constantly in contact throughout the game and tap out information to each other.

In essence, they were made to play footsie. As soon as the game had finished, all of the volunteers were asked to rate the attractiveness of the other players. The couple who had been behaving as if they were in love found each other more attractive.

Wegner was not the only one to try to conjure up love in the laboratory. In 2004, psychologists Arthur Aron and Barbara Fraley from Stony Brook University in New York carried out a similarly strange, but equally insightful, study into the topic using several blindfolds and a packet of straws.[26] Young lovers often have fun together, and so the researchers wondered whether encouraging people to behave as if they were having a good time in each other's company might pull them closer together. To find out, they invited participants to their laboratory, randomly paired them up, and then allocated them to one of two groups.

In one of the groups, the couples had a great time. The researchers blindfolded one member of the couple and had the other hold a drinking straw between his or her teeth, after which the researchers arranged for the blindfolded person to try to learn dance steps by listening to instructions read by the straw-holding colleague. After that, the blindfold and drinking straw were removed, and one member of the couple was given a pad and pen. The other was secretly given the name of a simple object, such as a tree or a house, and without naming it had to describe it while the partner attempted to draw the object. The couples in the other group weren't given the straws and blindfolds and were asked to learn the same dance and draw the object under more straightforward circumstances.

After the fun and games had stopped, all of the participants were asked to draw two overlapping circles to indicate the degree of closeness they felt with their partner. The pairs who had been behaving as if they were a happy couple felt significantly closer to each other.

Over the years, researchers have conducted many similar studies. In some experiments, psychologists have pretended to test for extrasensory perception and had people stare into each other's eyes, while in others, they have encouraged complete strangers to share their innermost secrets with one another. Time and again, the results have shown that it is possible to conjure up Cupid.[27]

Making Love

Encouraged by these findings, American psychologist Robert Epstein decided to take things one stage further. Was it possible, he wondered, to use these techniques to produce passion outside the laboratory?

Epstein has had a colorful career.[28] In his late teens, he felt a calling to become a rabbi and so sold his possessions and headed for

Israel. Six months later, he decided that he had misunderstood the nature of his calling and returned to America, determined to make a "significant and lasting contribution to mankind." He developed an interest in psychology and eventually enrolled as an undergraduate at Harvard. Publishing a remarkable twenty-one scientific papers in just four years, Epstein was excused from having to write a dissertation by the head of Harvard's psychology department, who advised him to "staple some of your publications together and get out while you still can." A few years later Epstein became the editor at the popular magazine *Psychology Today* (back issues are affectionately referred to by psychologists as "Psychology Yesterday"). In 2003 he left the magazine and has since been researching a range of topics, including creativity, stress, adolescence, and love.

Epstein is convinced that when it comes to love, the Western world is being sold a dangerous lie by fairy tales, romance writers, and Hollywood blockbusters.[29] From a very young age, children read stories in which damsels in distress are swept off their feet by knights in shining armor, and love is portrayed as a mystical emotion created by magical kisses, mysterious potions, or the will of gods. Later in life, adults read books and watch movies about people searching endlessly for "the one" and, if successful, living happily ever after. According to Epstein, these inaccurate conceptions of love seep into our minds and wreak havoc in our lives.

He believes that love is not a magical process and that people are not destined to be with one specific person. Instead, Epstein is convinced that love develops according to established psychological principles and that almost any two people could fall for each other by behaving as if they are in love.

The idea may sound outlandish, but there is some evidence that it might be true. Many celebrity couples have fallen for each other after acting out an on-screen romance. Richard Burton famously fell in love with Elizabeth Taylor when they were making the film *Cleopatra*. Brad Pitt and Angelina Jolie fell in love with each other while playing husband and wife in *Mr. & Mrs. Smith*. And in *Bugsy*,

Warren Beatty plays the role of the gangster Benjamin "Bugsy" Siegel who instantly falls for a Hollywood starlet played by Annette Bening. Beatty and Bening married soon after filming. In each instance, the celebrities acted as if they were in love and promptly fell for each other for real.

In June 2002 Epstein, then in his late forties and single, announced that he intended to carry out a "bold, very personal" study to discover whether his theory about love was correct.[30] Writing in *Psychology Today*, he described how he would try to find a woman who was willing to join him in an experiment to discover if it were possible for two strangers to learn to love each other. Rather than go through the horrors of dating, Epstein and the selected woman would spend between six and twelve months following a simple set of rules designed to bring them together (such as agreeing not to date others and to participate in exercises designed to promote love), and then coauthor a book about their experiences (working title: "The Love You Make: How We Learned to Love Each Other, and How You Can Too"). Explaining that the idea was not just a publicity stunt but a serious study into the nature of love, Epstein said that several large publishing houses had already shown an interest in the book.

The idea was quickly picked up by the media and more than a thousand women volunteered for the experiment. Epstein met fifteen of the applicants but rejected them all, later explaining that many of them appeared more interested in obtaining publicity than genuinely learning to love him.[31] Then, on Christmas Day 2002 Epstein was on a flight and found himself seated next to a Venezuelan ex-ballerina named Gabriela Castillo. The two of them started to chat, and Epstein eventually explained his experiment and asked Castillo whether she would like to be "the one." At first Castillo was reluctant to get involved but eventually agreed, and on Valentine's Day 2002 Castillo and Epstein signed the "love contract." Unfortunately, the couple struggled with the long-distance nature of their relationship (Castillo was based in Venezuela, Epstein in America), and despite several visits to relationship counselors, Castillo and Epstein eventually decided to abandon the experiment after a few months. In 2008 Epstein married a woman he met after giving a lecture on the Isle of Man.[32]

Unperturbed by the failure of his personal love project, Epstein has since created a series of exercises designed to promote love outside the laboratory and tested them on students at the University of California, San Diego. These exercises encouraged pairs of strangers to carry out a variety of love-inducing tasks, including gently embracing each other, synchronizing their breathing, gazing longingly into each other's eyes, falling into each other's arms, and getting physically close to each other without touching (Epstein reports that this last exercise apparently often ends with kissing).

Epstein asked his romantic guinea pigs to rate the degree to which they felt emotionally close to one another before and after the exercises, with the results showing that afterward, the couples did find each other attractive and did move emotionally closer to each other.[33]

The results sounded encouraging. Could this approach to love help those genuinely searching for Cupid's arrow? It was time to find out.

Moving On

Struggling to get over a relationship? The As If principle can help.

Researcher Xiuping Li from the National University of Singapore Business School asked eighty people to write down a recent decision they regretted. Li then asked some of the participants to hand their descriptions to a researcher and others to seal it in an envelope. The participants who sealed their experience in an envelope felt significantly better about their past decision compared with those who handed it over. Sealing the description in an envelope meant that the participants were acting as if they had reached psychological closure and were moving on.

Next time you want some help getting over the end of a relationship, write a brief description of what happened on a piece of paper, rip up the paper, put it in an envelope, and kiss the past good-bye.

And if you really want to have fun, reach for the matches and convert your envelope into a pile of ashes.

Speed Dating in the Fast Lane

Speed dating can often be a dull and repetitive process, with people talking about the same topics time and again throughout the evening. I wondered whether it might be possible to use the As If principle to create a new form of speed dating that was more interesting and effective.

I rented a gorgeous Georgian ballroom in the center of Edinburgh and advertised for single people who wanted to take part in a study exploring the science of seduction. I then invited twenty single men and twenty single women to my love laboratory.

Before the start of the evening, we placed candles on each of the tables, turned the lights down low, and piped in romantic music. The scene was set. As each of our participants arrived, they were seated at a long table, men on one side and women on the other. Everyone was handed a "book of love" containing the instructions for the evening (see "The Book of Love" box).

Once everyone was sitting comfortably, we began. In the first exercise, everyone was asked to chat with the person opposite them and find out the person's name and background. Then we handed everyone a blank cardboard badge and some marker pens and asked everyone to make a badge for their partner. The badge had to incorporate the partner's name and something interesting about him or her. Finally, we asked everyone to give the badge that they had just made to the partner. Loving couples often make and give little gifts to each other, and this exercise was designed to get our participants to behave as if they found each other attractive.

At the end of this first exercise everyone was asked to check one of two boxes to indicate whether they would like to see their partner after the speed-dating evening. All of the women then swapped seats, moved on to their next partner, and carried out the second exercise. This process carried on throughout the entire evening, with each exercise involving a different behavior and a different partner. Some of the time the participants were asked to gaze into each other's eyes or hold hands, at others they swapped secrets or worked together to achieve a single goal.

The Book of Love

Here are some of the most successful love games used during the experimental speed-dating study. With a little ingenuity, they can be used to promote attraction in everyday life.

Reading Minds (Eye Contact)
Secretly draw a simple picture on a piece of paper. Next, spend forty-five seconds staring into your partner's eyes trying to "transmit" the drawing to him or her telepathically. Ask your partner to draw the picture that he thinks you were trying to send him on another piece of paper. Finally, compare the two pictures. Spend a few minutes discussing

whether the pictures match, why you drew your picture, and why your partner drew his or her picture.

Secrets (Sharing Secrets)

Both you and your partner discuss the following five questions:

1. Name something that you have always wanted to do, and explain why you haven't done it yet.
2. Imagine that your house or apartment caught fire. You can save only one object.
3. What would it be?
4. What advice would you give your ten-year-old self?
5. What do you like best about your life?
6. When was the last time that you cried with laughter?

Knowing Me, Knowing You (Exploration)

Take turns answering the following five questions:

1. If you had a superpower, what would it be?
2. Which celebrity would you most like to go to dinner with?
3. If you travel back in time, what era would you visit?
4. If you could have any job in the world, what would it be?
5. If you won the lottery tomorrow, how would you spend the money?

Everyone had fun staring into each other's eyes and discussing their innermost secrets, but did the exercises help promote that loving feeling? I have run several other conventional speed-dating sessions over the years, and around 20 percent of the encounters result in both participants' checking the "yes, I would like to see this person again" box. When the As If principle was at work, the success rate was a remarkable 45 percent. Just spending a few moments behaving as if they found each other attractive was enough to help get in touch with their inner Cupid.

At the start of the evening, Lianne and Nick arrived as singletons.* When they were paired up, they were asked to take part in a game designed to get couples talking about their lives and touching each other in an acceptable way and to have fun together. During the game Lianne and Nick made lots of eye contact and enjoyed the excuse to hold each other's hands. Their chat after the game revealed that the two of them had lots in common and made each other laugh.

Both of them indicated that they would like to see each other again, and so I arranged for an email introduction. They arranged to meet for coffee the following week. Once again, everything went very well, and the coffee turned into dinner and a couple of glasses of wine. After meeting again a few days later, Lianne and Nick fell for each other and are now in a loving relationship.

The As If principle can clearly help people get together, but can it make them live happily ever after?

Creating a Happy Ending

When people first fall in love, they tend to go out and about together and try lots of new and exciting experiences. However, as time goes on, it is all too easy for couples to get stuck in a rut. Finding themselves having the same conversations and visiting the same places time and again, they can become bored with each other's company. Indeed, several research projects have discovered that boredom is one of the main sources of an unhappy marriage.[34] Psychologist Arthur Aron (he of the shaky bridge, blindfolds, and straws fame) wondered whether getting long-term couples to behave as if life was fun again would make them feel more loving toward each other.

---

* Lianne and Nick are pseudonyms.

Aron recruited fifty married couples who had been together for an average of fourteen years and persuaded them to take part in a ten-week experiment.[35] He presented everyone with a long list of activities and asked them to rate how enjoyable and exciting they found each of the items. Next, he split the couples into two groups, and asked the couples in one group to spend one and a half hours each week carrying out an activity that they found enjoyable and the couples in the other group to spend the same amount of time on an activity that they found exciting.

At the end of the study Aron asked everyone to rate how happy they were with their marriage. Those who had spent time engaging in exciting activities (such as skiing, hiking, dancing, or going to concerts) were significantly happier with their relationships than those who had been encouraged to carry out pleasant activities (such as going to the movies, eating out, or visiting friends).

His results show that the key to long-term love involves people avoiding the lure of the familiar and instead inviting excitement into their lives. By acting as if they are out on an exhilarating date, couples can turn back the hands of time and easily recreate that loving feeling.

The Dice Man

This exercise is designed to help existing couples bring back the magic of their first few years together. You and your partner should each complete part one of the exercise separately.

Part One: Take a look at each of the following activities and make a note of the ones that you find especially exciting.

Go for a walk in the countryside	Go for a ride in a speedboat
Go to see a live concert	Eat snails
Play a sport	Fly a kite
Plan a trip or vacation	Go on a road trip
Go on a shopping trip	Bet on a horse race
Go to the beach	Kiss on a fairground ride
Create some kind of artwork	Place a pin in a map and go
Rearrange or redecorate your	to that place
home	Enter a trivia quiz in a bar
Go to a sports event	Learn some circus skills
Go to a new restaurant	Go canoeing
Go to a lecture or talk	Arm-wrestle
Go camping, hiking, or boating	Dive or jump off a high
Invite friends to dinner	board
Learn to windsurf	Sleep out under the stars
Go dancing	Fly in a seaplane
Visit a fair or zoo	Go on a long train journey
Get a massage or go to a health	Ride on a huge roller-
club	coaster
Plan a major purchase	Travel on a hot-air balloon
Visit a museum or art exhibition	Swim with some dolphins
Go to a movie	Do a parachute jump

Now write down two other activities that you find exciting.

Part Two: Sit down with your partner and look at your ratings and answers. Select six activities that both you and your partner find exciting, write them on a piece of paper, and number them one to six.

In the early 1970s, author Luke Rhinehart published a novel entitled *The Dice Man.* The book tells the story of a psychiatrist who starts to make major decisions based on the roll of a die. It is time for you to play the role of the dice man. Find a die, roll it, and read out the corresponding chosen activity. Ensure that you carry out this activity during the next two weeks and repeat this process every two weeks.

For centuries scientists have struggled to understand the mysteries of love. According to the small number of conventional theories that have been developed, love makes your heart beat faster and motivates you to stare longingly into your partner's eyes. The As If principle shows that the exact opposite is true: behaving as if you are in love can ignite the flames of passion. Encourage complete strangers to hold hands and play footsie, and suddenly Cupid loads his bow ready for action. Have long-term couples behave as if they are reliving the excitement of their first date, and suddenly they find each other irresistible all over again. This simple but important principle can help people find love and live happily ever after.

It is not that love changes everything, but that changing your behavior can create the world's most desirable emotion.

Chapter 3
Mental Health

Where we meet the "Napoleon of neuroses," find out why watching sports is bad for your health, and discover how best to deal with phobias, anxiety, and depression

"Action is the antidote to despair."

—Joan Baez

I. Paralysis and Emotion
II. Eliminating Pain, Anger, and Anxiety
III. Dealing with Depression

I. Paralysis and Emotion

As you read this, millions of people across the world are struggling with some form of psychological disorder. Some will be trying to live with an irrational phobia, others will be overly anxious, and many will be depressed. For more than a century, scientists and psychologists have attempted to cure these problems. They have adopted many different perspectives, including giving people drugs, operating on their brains, and talking to them. Can the As If principle help people to move away from the dark side?

William James first outlined his radical theory in an essay entitled, "What Is an Emotion?" The essay ended with a brave and bold prediction: if behavior created feelings, then people who were totally paralyzed shouldn't be able to experience emotion. James also realized that it would be tricky to carry out this test, in part because of the difficulties in determining the emotional life of someone who was completely immobile ("It must be confessed that a crucial test of the truth of the hypothesis is quite as hard to obtain as its decisive refutation"). Eighty years after James wrote his now-classic essay, researchers devised a clever way of carrying out a version of his test and, in doing so, laid the foundations for a new approach to pain management, phobias, anxiety, and depression.

To appreciate their remarkable study, it is important to understand what is happening inside your body. You are amazing. Really, you are. At this very moment, millions of electrical impulses are zipping around a biological superhighway that covers you from head to toe.

One lane of this highway is carrying information from your sensory receptors to your brain. These receptors will be hard at work as you read these sentences. If you are sitting down, the receptors in your legs and buttocks will be continuously sending "I am being squashed by the weight of my upper body" signals to your brain. Each time you turn a page (or, if you are using one of those newfangled ebook readers, whenever you press the forward key), the receptors in your fingertips will send information about their movements up to your head. Your bladder and digestive system are also using the superhighway to tell your brain whether you need to go to the bathroom or get something to eat. Similarly, the million or so fibers that continually carry information from the back of your eye to your brain will be buzzing with excitement, busily communicating the shape of each letter on the page and delighted to get a name check in such a prestigious book.

The other lane in the biological superhighway carries information in the opposite direction, transmitting signals from your brain to your body. If you are sitting down at the moment, these impulses will be continually sending small signals to your major muscles to ensure that you remain balanced on the chair. Each time you turn a page, or press the forward button, the signals control the delicate actions your hands and fingers make. When your brain gets excited, it uses the highway to send signals that speed up your heart and increase your rate of breathing. And as you read these words, other signals on the highway will be rapidly scanning each of the "the" lines on the page, slightly annoyed that they just missed the two "the's" in the middle of this sentence.

The superhighway starts at the base of your brain, runs down your spine, and is peppered with junctions that allow information to travel to and from various parts of your body. Junctions toward the base of your spine take information to and from your feet and legs, those toward the middle of the highway carry impulses to and from your hands and arms, and those toward the top of the highway communicate with your face and eyes. Similarly, information to and from your bladder joins the highway toward the bottom of your spine, your digestive organs join slightly farther up, and your heart is controlled by signals coming from the junctions toward the top of the spine.

If this entire complex system of interconnected signals were to suddenly fail, you would experience multiple organ failure and die in an instant. However, the good news is that for most people, the system works extremely well. Each moment of your life, information zips around the highways and byways of your body and brain, allowing you to perceive your surroundings, move around, and stay alive. Not only that, but the entire system has evolved to operate outside conscious control, allowing you to focus on the finer things in life, such as appreciating great art, trying to understand scientific advances, and finding a plumber on a Sunday.

In the mid-1960s psychologist George Hohmann was working at a Veterans Administration hospital in Arizona. Hohmann was a paraplegic and realized that his patients provided a unique opportunity to test James's "immobility will prevent emotion" prediction.[1] Many of Hohmann's patients were paralyzed because they had suffered an injury to their spinal cord, and the level at which the injury had occurred was directly related to their degree of immobility. An injury toward the base of the spine, for example, would have severed the lower part of the superhighway and so resulted in a loss of movement and feeling in the legs. In contrast, an injury higher up the spine would have disrupted a much larger part of the highway and so resulted in a loss of feeling and movement in both the legs and arms. Hohmann reasoned that if James were correct, the higher the lesion on the spine (and therefore the less bodily movement), the greater the loss in emotional experience.

Hohmann tracked down patients who had received an injury to one of five sections of the spine and interviewed them about their emotional experiences. In one part of his study, Hohmann asked his patients to compare how often they were afraid before and after their injury. Patients with injuries to the lower parts of their spine reported very little difference, while those with lesions toward the top of the spine reported becoming fearless. The patients' reports provide a remarkable insight into life without emotion, with one fearless patient noting, "Sometimes I act angry when I see some injustice. I yell and cuss and raise hell, because if you don't do it sometimes, I've learned people will take advantage of you, but it just doesn't have the heat to it that it used to. It's a mental kind of anger."

When Hohmann asked the patients about other emotions, such as grief, the same pattern emerged: the higher the spinal damage, the less able they were to move around and the less likely they were to experience emotion.

Hohmann's findings are a remarkable tribute to James's genius and demonstrate the vital role that your body plays in determining how you feel. Exactly as James predicted over eighty years before, the higher the damaged area is on the spine, the more dramatic the drop is in emotional experience.[2]

More recently, other researchers wanted to discover whether James's idea is also true for facial expressions. Would people who are unable to move their faces show a drop in emotional experiences? The researchers could have spent years tracking down people with different levels of facial paralysis and then assessing their emotional experiences. Instead, they adopted a lateral approach to the problem and saved themselves a great deal of time and trouble, by working with an easily available group of people who had voluntarily paralyzed their own faces.

Botox (or, as it is known in scientific circles, botulinum toxin) is one of the most popular cosmetic treatments in the world. Originally developed to help people suffering from facial muscle spasms, Botox paralyzes the nerves that cause the facial muscles to contract. In the early 1990s, researchers discovered that injecting the chemical into the frown lines between the eyes caused partial paralysis of the forehead and so significantly reduced wrinkling. Although this can result in a more youthful appearance, it can also sometimes make people's faces appear a tad expressionless and robotic.

Joshua Ian Davis from Barnard College and his colleagues wondered whether such attempts to turn back time could form the basis of a remarkable test of James's theory.[3] He recruited two groups of women for the experiment. One group had just undergone treatment with Botox injections, while the other group had opted for a different treatment that involved injecting a kind of filler into the forehead. Both treatments aimed to create a more youthful appearance, but only the Botox paralyzed the facial muscles. Davis asked the women to watch several video clips, including a scary clip of a man eating live worms, a hilarious clip from America's funniest videos, and a serious documentary about Jackson Pollock. After seeing each of the clips, the women were asked to rate how they felt. Compared with those who had had the filler treatment, the Botox women reported less of an emotional reaction to the clips. The evidence is that James was right: immobility (in this instance of facial expressions) causes a loss in emotional experience.

The message from the patients with spinal injuries and the women injecting Botox is clear: inhibiting people's behavior and facial expressions prevents them from feeling certain emotions. On the downside, such individuals are less likely to experience positive emotions such as happiness and joy. However, on the upside, they are also less likely to feel negative emotions, such as anger and anxiety. Researchers became curious and began to examine whether this latter result could be used to help avoid unwelcome feelings.

II. Eliminating Pain, Anger, and Anxiety

In the 1970s British doctor Peter Brown visited a children's hospital in China to observe how doctors there carried out tonsillectomies.[4] Brown was stunned by what he saw.

Patients undergoing tonsillectomies in the West often reported experiencing a great deal of pain. In China, the situation was quite different. Brown reported seeing a line of smiling five-year-olds standing outside the doctor's surgical unit. Each child was given a quick anesthetic throat spray by a nurse and then led into the room. Once inside, the still-smiling child climbed onto a table and opened his or her mouth. Within seconds the doctor whipped out the tonsils and dropped them into a bucket of water. The child walked to a recovery room showing few signs of discomfort.

The vast difference in discomfort by patients undergoing tonsillectomies in the West and China illustrates the subjective nature of pain. It isn't an isolated example. Often people experience exactly the same hospital operation, event, illness, or accident but report very different levels of pain. Why should this be the case? According to the As If principle, a large part of the answer has to do with different ways in which they behave.

Some social psychology experiments involve telling participants that they are administering dangerous electric shocks to another person when, in reality, the "shocks" are harmless and the other person is a stooge. However, before the creation of university ethics committees (an era that many psychologists nostalgically refer to as "the good old days"), some studies did actually administer genuine and painful shocks to volunteers.

One such study was conducted by John Lanzetta and his colleagues from Dartmouth College.[5] Lanzetta invited volunteers to his laboratory one at a time and connected them to two machines. First, the experimenters placed electrodes on the volunteer's legs and left hand and plugged the wires into an electric shock generator. Then they attached sweat sensors to the volunteer's right hand in order to be able to continuously measure the person's stress levels. After ensuring that the machines were ready to shock and sense, the experimenters retreated to an adjoining room.

A closed-circuit television system ensured that the experimenters could see, talk to, and hear the participant. They told the volunteer that he or she was about to receive a series of electric shocks of varying intensity, and that the participant was to rate the pain associated with each shock by calling out a number between 1 ("whatever") to a 100 ("just you wait until you come back in here"). The experimenters then administered twenty electric shocks and carefully noted down the numbers each volunteer screamed out.

After a short break, Lanzetta explained that there was going to be a second series of shocks, but that this time the volunteers should do their best to hide how they felt. Each volunteer was asked to play tough, squash any emotional expressions, avoid screaming, and adopt a relaxed posture. A second round of twenty shocks was then administered, and each time the participants shouted out their ratings.

The results were quite remarkable: when the volunteers acted as if they were not in any discomfort, they experienced much lower levels of pain. Not only that, but the data from the sweat sensors revealed that they were indeed far less stressed. The study has been repeated many times over the years, and the same results have been obtained.

This rather counterintuitive effect helps explain why the Chinese children appeared so unperturbed during the tonsillectomies. At the time of Peter Brown's visit, Chinese children were taught to view operations in a positive way and so would often smile and behave in a relaxed way when undergoing medical procedures.

The effect also helps explain several other curious pain-related phenomena. For example, people experience less pain during minor medical procedures if they look away from the incision or injection. In doing so, they are far less likely to adopt a pained facial expression or tense up and so feel less discomfort. The same goes for other effective pain-reduction techniques that involve distraction, such as imagery, hypnosis, and relaxation techniques. Each time people are behaving as if they are not in any discomfort, and this causes them to experience less pain.

Inspired by this effect, researchers then examined whether behaving as if you are a strong and powerful person would result in even lower levels of pain. Vanessa Bohns from the University of Toronto and colleagues told a group of volunteers that they were taking part in a study about the health benefits of exercise at work.[6] Some of the volunteers were asked to adopt a posture associated with dominance and power by puffing out their chests and moving their arms away from their body. In contrast, others were encouraged to curl up in a powerless-looking ball. Next, the experimenters placed a tourniquet around each volunteer's arm and slowly inflated it. The band slowly reduced blood flow and so became increasingly painful, and volunteers were asked to say when they could no longer tolerate the discomfort. Remarkably, those in the powerful posture were able to tolerate much tighter tourniquets than those curled up in a ball. Behaving as if they were powerful and strong helped push away an unwanted emotion and showed that the age-old expression, "Keep your chin up," may be literally true.

Results from the early work into the As If principle and pain motivated researchers to examine whether the same idea could be used to help minimize other unwanted emotions. Could it, for example, help pour cold water on those seeing red?

You Wouldn't Like Me When I Am Angry

Anger is bad for you. It often makes you act foolishly, take irrational risks, blurt out comments that you later regret, and lash out (the vast majority of murders committed in the United States are due at least in part to the effects of anger).[7] It is also bad for those around you. Psychologist Martin Seligman tracked the lives of four hundred children for five years, focusing on families in which the parents routinely fought with each other.[8] Seligman's findings show that the children from warring families were more likely to be diagnosed with depression and struggle later in life.

What is the best way of keeping your inner Hulk under control? To find out, we have to travel back to the turn of the last century and spend some quality time with one of the world's most famous psychologists, Jean-Martin Charcot, described as "the Napoleon of the neuroses." A charismatic lecturer with a flair for the dramatic, this nineteenth-century French clinician paved the way for modern neurology, has more than fifteen diseases named in his honor, and carried out groundbreaking research into the causes of both multiple sclerosis and Parkinson's disease. Despite such impressive achievements, Charcot is best known for his remarkable journeys into the unconscious mind.

Charcot was fascinated by the secret life of the brain. To carry out his investigations, he frequently teamed up with patients from Parisian insane asylums and embarked on bizarre explorations into the unconscious. Much of the work was carried out live as Charcot lectured to fellow doctors. In 1887 the French artist André Brouillet attended several of Charcot's lectures and painted the Napoleon of neuroses in action.[9] The image shows Charcot standing on the right-hand side of the painting wearing a fashionable black suit. On the left of the scene, a group of around thirty men are carefully watching Charcot and taking notes. An unconscious woman is draped across Charcot's outstretched left arm.

The woman in Brouillet's painting is one of Charcot's star performers, Blanche Wittmann. Described by historians at the time as having "a large and strong body" with "very large bosoms," Wittmann had been admitted to an asylum after having several attacks of hysteria, frequently urinating on herself, and admitting to sexual intercourse with her employer. During his lectures, Charcot would place Wittmann into a hypnotic trance and then ask her to exhibit a range of strange phenomena, including falling into a catatonic state, maintaining an odd arched position by balancing on top of her head and the points of her feet, performing mirror writing, and making selected words appear scratched on her skin. Charcot argued that these phenomena were manifestations of Wittmann's unconscious mind and so could be used to explore the innermost workings of the brain. At the end of each session, Charcot would compress Wittmann's "ovarian region" to awaken her from the trance, whereupon she would return to the land of the living with her pupils "greatly dilated."

Charcot's dramatic approach to the human psyche quickly became the talk of the town, and academics from across Europe traveled to see his curious demonstrations. In 1885 a twenty-nine-year-old Austrian physician named Sigmund Freud attended one of Charcot's demonstrations. Before seeing Charcot in action, Freud had intended to pursue a career in medicine and had previously carried out a considerable amount of research in which he had dissected hundreds of eels in an unsuccessful search for their reproductive organs. After seeing Charcot dilate the pupils of several young female patients, Freud became convinced that the unconscious played a key role in a number of psychological disorders.

After taking vast amounts of cocaine and with a cigar rarely out of his mouth (make of that what you will), Freud eventually came up with a dramatically new form of psychology known as psychoanalysis. According to Freud, people tend to push unwanted thoughts out of their conscious awareness and into the unconscious. Once there, the threatening thoughts bubble away and build up psychic energy. When they have gained sufficient psychological strength, the thoughts start to affect consciousness in a variety of unhealthy ways, creating, for example, feelings of unease, neurosis, and anxiety.

Freud believed that it was psychologically healthy to get rid of these repressed thoughts before they became too powerful, and he attempted to develop therapeutic techniques that could act as a release valve for the unconscious mind. Initially following in the footsteps of Charcot, Freud started off by trying to hypnotize his patients. When this had little effect, Freud promptly abandoned the method and instead explored several other approaches, including dream analysis (where the therapist attempts to reveal the symbolic meaning of patients' dreams) and free association (where the patient is asked to say the first word that enters his or her head when the therapist says carefully chosen stimulus words such as *chair, table,* and *orgasm*). Applying these techniques to his own unconscious, Freud became convinced that he had developed sexual feelings toward his mother when he was around two years old.

Freud's ideas quickly caught on, and by the turn of the last century, psychoanalysis was starting to spread across the world. In 1909 Freud was invited to Clark University in Massachusetts to give a series of prestigious lectures. It was the only time that he spoke in America, and he used the opportunity to deliver an overview of his beloved psychoanalysis.

At the time of Freud's visit, William James was sixty-seven years old and suffering from a painful heart condition. Despite his poor health, he traveled to Clark University to hear Freud talk.[10] James was less than impressed, later describing Freud's idea of dream symbolism as "a dangerous method" and believing that the great psychoanalyst himself was both deluded and "obsessed by fixed ideas."

James and Freud differed on many topics, including the causes and treatment of excessive anger. According to Freud, people became angry because they were repressing violent thoughts, and it would therefore be cathartic for them to release these feelings in a safe way by, for example, punching a pillow, shouting and screaming, or stamping their feet. In contrast, James's theory predicted that people become angry because they are behaving in an angry way and that Freud's cathartic therapy would simply make them angrier. For years psychologists have carried out studies to discover which of the two great thinkers was right.

One of the first into the fray was sociologist Murray Straus from the University of New Hampshire.[11] In the early 1970s, Straus became concerned with what he saw as the overly Freudian advice being given to couples struggling to maintain their relationship. Much of this advice came from the therapeutic aggression movement, which thought that couples shouldn't hold back when it came to telling each other what they thought. Manuals from the time encourage couples to "Get rid of your pent-up hostilities," "Let it be totally vicious," and to bite a plastic baby's bottle while imagining that it is their partner.

To find out if such radical behavior helped or hindered a relationship, Straus carried out a simple study. He reasoned that if the cathartic theory were valid, couples that are verbally aggressive should be less likely to carry out acts of actual physical aggression toward each other. Straus also realized that couples might not be the best ones to report accurately on their aggressive behavior and so asked his students to secretly monitor their parents' level of verbal and physical aggression. More than three hundred students carefully completed checklists about how their parents responded when they were faced with a problem. Did they tend to discuss the issue in a constructive way? Did they become verbally aggressive, perhaps yelling and storming out of the room? Did they become physically aggressive, perhaps throwing objects at each other or hitting their partner?

When Straus analyzed the results, an obvious pattern emerged: the more the couples engaged in verbal aggression, the more likely they were to be physically aggressive. Just as James predicted, shouting and yelling wasn't cathartic; rather it encouraged people to behave in an angry way. Freud 0, James 1.

Next were studies carried out in the workplace. Ebbe Ebbesen from the University of California, San Diego, and his colleagues discovered that a local engineering firm was going to lay off many of its employees.[12] The employees had every right to be angry because they had been promised a three-year contract but were being put out of work after just one year. Ebbesen interviewed some of the employees in one of two ways. One group was encouraged to talk about how angry they were at the firm ("So, how do you feel about the way in which you have been treated?"), while the other group was asked much more neutral questions ("So, can you describe the firm's technical library?"). After the interviews, all of the employees rated their level of hostility and anger toward the company. Did those who had been encouraged to rant and rave show lower levels of hostility? No. Once again, the exact opposite pattern of results emerged: the employees who had vented their feelings of anger were much more hostile than those who had been asked about the company's technical library. Freud 0, James 2.

Finally, there is the work investigating the link between hostility and watching sports. When people attend football games, they often cheer on their team and shout abuse at the opposition. Freudians had argued that such aggressive behavior was cathartic, so people would feel significantly less hostile after the game. In contrast, proponents of James's ideas thought that all of the shouting and jeering would make people feel far angrier. Jeffrey Goldstein from Temple University decided to discover which camp was correct.[13]

Goldstein arranged for a small army of researchers to attend a major American football game. Before the start of the game, the researchers hung around the stadium turnstiles and interviewed a randomly selected group of spectators. The interviews were short and included asking each spectator which team he or she supported and how aggressive he or she felt. Once the game had ended, the researchers took to the turnstiles again and interviewed randomly selected spectators as they left the game.

The results revealed that regardless of whether the spectator's team won or lost, the spectators were significantly more aggressive after the game. Worried that this increase in aggression might be due to spectators being in a crowd or simply watching a competitive situation, Goldstein reassembled his crack research team and had them carry out the same interviews at a local gymnastics competition. Although the spectators attending the gymnastics event were crowded together in a competitive context, they didn't shout and jeer and weren't significantly more aggressive after the event. Goldstein's data suggest that the football game encouraged the spectators to behave in an aggressive way and that this behavior made them feel more hostile. Freud 0, James 3.

Such induced hostility can have a very real impact on society. Glasgow in Scotland has two professional football clubs. Celtic is based in the east of Glasgow and has traditionally had support from the Catholic community, while Rangers has its base in the southwest of the city and has historically attracted a Protestant following. The clubs have a long-standing and fierce rivalry, with supporters from both sides often singing hostile and intimidating chants during matches. In 2011, researchers working for the Scottish police compared the levels of reported crime after Rangers-Celtic matches with times when the teams were not playing.[14] The figures are remarkable: when Rangers played Celtic on Saturday lunchtimes, violent crime in Glasgow almost tripled and incidents of domestic abuse more than doubled.

The Power of Calm

Psychologist Brad Bushman from Iowa State University has carried out several experiments showing how feelings of anger can be quickly squashed by acting like a calm person. In one study, for example, Bushman had college students spend twenty minutes playing either a relaxing or a violent computer game.[15] In one of the relaxing computer games, the students swam around a quiet underwater habitat looking for sunken treasure, while in the violent games, they tried to dispatch as many zombies as gorily as possible. The students were then asked to play another game against an unseen player; if they won, they were given the opportunity to blast their opponent with a loud noise. In reality, there was no other unseen player, and the students always won this second game. Those who had previously spent some time quietly swimming around the underwater habitat were significantly less aggressive, choosing a quieter and shorter noise blast for their imaginary opponent.

Bushman has also demonstrated the calming power of prayer.[16] He angered a group of Christian college students by giving them extremely negative feedback about their work ("Christ, this is one of the worst essays I have read") and then asked them to read a newspaper article about a woman with a rare form of cancer. Next, some of the students were asked to spend five minutes putting their hands together and praying for the woman, while the others were asked to think about her. Afterward, the students who had prayed were significantly less angry than those who thought about the woman. Acting in a relaxed and calm way produced relaxed and calming thoughts.

Also, in my previous book, *59 Seconds,* I described how Bushman used the same essay-based anger-inducing procedure to enrage another group of students.[17] Some of these students were then given a pair of boxing gloves, shown a photograph of the person who had allegedly graded their essay, and told to think about that person while they hit a seventy-pound punching bag. Another group was asked to sit in a quiet room for two minutes. The findings came down firmly against Freud: punching the bag made people feel angrier, while quietly sitting down made them feel much calmer.

Many anger management courses teach people to work hostility out of their system by behaving aggressively. This is unhelpful and, if anything, will tend to make the problem worse. Other approaches are trying to figure out the deep-seated psychological root of the anger in the hope that changing the way you think will change how you feel. In reality, there is a much quicker and more effective way of dealing with the problem. To calm down, act like a calm person (see the "Calming Down" box). In the same way that smiling will make you feel happy, and gazing into the eyes of another person will make you feel as if you are in love, acting in a calm fashion will quickly make you feel calm.

Calming Down?*

Those needing to deal with anger in a quick and effective way often benefit from a deep-breathing exercise. To try the technique, put your tongue on the roof of your mouth just behind your front teeth. Now breathe in slowly through your nose for the count of five and hold your breath for the count of seven. Then exhale slowly through pursed lips for the count of eight. Repeat this exercise four times.

For a more long-lasting solution others learn how to carry out progressive muscle relaxation. This approach involves deliberately tensing various muscle groups and then releasing the tension. To try this technique, remove your shoes, loosen any tight clothing, and sit in a comfortable chair in a quiet room. Focus your attention on your right foot. Gently inhale and clench the muscles in your foot as hard as possible for about five seconds. Then exhale and release all of the tension, allowing the muscles to become loose and limp. Work your way around your body performing the procedure in the following order:

1. Right foot
2. Right lower leg
3. Entire right leg
4. Left foot
5. Left lower leg
6. Entire left leg
7. Right hand
8. Right forearm
9. Entire right arm
10. Left hand
11. Left forearm
12. Entire left arm
13. Abdomen
14. Chest
15. Neck and shoulders
16. Face

---

* The exercises described here are designed to provide general insight into the sorts of techniques that psychologists use. If you believe that you have an anger management issue, consult a professional.

Little Hans and Albert B.

John B. Watson changed the entire course of psychology and shaped our current understanding of the human psyche.[18] Working at Johns Hopkins University just after the turn of the twentieth century, Watson by any measure was a strange and complex man. To outsiders, he appeared flamboyant, outgoing, and confident. In reality, he was deeply insecure, petrified of the dark, and emotionally frigid. A tad socially awkward, Watson would shake hands with his children rather than kiss them goodnight and leave the room whenever anyone tried to discuss his emotions.

Watson firmly rejected both Wilhelm Wundt's introspection and Sigmund Freud's psychoanalysis. Arguing that as it was impossible to know for certain what was really going on in people's minds, he believed that psychologists should instead focus their attention on observing and measuring behavior (thus giving rise to the joke wherein two behaviorists make love, and then one turns to the other and says, "That was good for you. How was it for me?").

Watson loved to run rats through mazes. In his early years, he constructed a miniature version of the medieval maze at London's Hampton Court Palace, placed hungry rodents in it one at a time, and carefully observed them running around trying to find a strategically placed pile of food. After testing hundreds of rats in this real-life version of, "Who moved my cheese?" Watson worked out the rodent rules of basic learning, including how the rats explored the maze and how long the rats would continue to visit a location that used to contain food even after the food had been removed.

Watson became convinced that his results applied to human behavior and that the whole of life was like a huge maze. Even more controversial, Watson thought that it was possible to shape any human mind by simply applying the rodent rules of learning that he had uncovered during his maze experiments, causing him to famously remark:

> Give me a dozen healthy infants, well-formed, and my own
> specified world to bring them up in and I'll guarantee to take any

one at random and train him to become any type of specialist I might select—doctor, lawyer, artist, merchant-chief and, yes, even beggar-man and thief, regardless of his talents, penchants, tendencies, abilities, vocations, and race of his ancestors.[19]

Watson's total focus on the measurement of behavior quickly caught on, and soon researchers across the world started to run more and more rats through increasingly complex mazes, causing one commentator to remark, "Psychology, having first lost its soul to Darwin, has now lost its mind to Watson."

The behaviorists started to expand their empire away from the rules of learning and into other areas of psychology, with Watson taking a particular interest in the cause and treatment of phobias. Like many other behaviorists, he was driven by a desire to provide an alternative to what he saw as the pseudoscientific ramblings of Sigmund Freud.

Freud encouraged an inner circle of supporters to help develop psychoanalytical theories by secretly observing the sexual life of their young children. In 1904 one of Freud's closest associates reported that his five-year-old son, referred to as "Little Hans," had developed an irrational fear of horses and suggested that this might make for an interesting case study. Freud agreed and began to explore what might be causing the young boy to fear horses. Hans's father initially attributed the phobia to Little Hans's sexual overexcitement when his mother cuddled him, coupled with his shock by the large penises of horses. Freud disagreed and instead noted that the boy had described dreaming about a giraffe and that the long neck of the giraffe was a symbol for the large adult penis. After much back and forth, Freud described his thoughts in a paper, "Analysis of a Phobia in a Five-Year-Old Boy," and suggested that Little Hans's disorder was the result of several factors, including a repressed desire to make love to his mother and conflicting thoughts about masturbation.

Watson was appalled by Freud's speculations about Little Hans's alleged emotional inner turmoil and became determined to come up with a far more down-to-earth explanation of phobias. Watson's approach was heavily influenced by the work of the Russian researcher Ivan Pavlov. A few years before Watson had started running rats through mazes, Pavlov had been observing the impact of ringing bells on dogs. Pavlov had carried out a now-classic set of studies in which he had rung a bell and then presented a dog with a bowl of food. Not surprisingly, the dogs had salivated at the sight of the food. After just a handful of these trials, Pavlov discovered that just ringing the bell was enough to cause the dogs to salivate and so demonstrated that brains are extremely skilled at learning such associations (a procedure known in the trade as classical conditioning).

Pavlov's simple but important discovery has several practical applications. In one study, for example, animal researchers laced some sheep carcasses with a poison that made coyotes vomit and put the carcasses out in a field. In the same way that pairing the bell with the presentation of food made Pavlov's dog salivate, so wolfing down the poisoned sheep made the coyotes feel ill whenever they saw a sheep. As a result, the number of coyote attacks on sheep dramatically declined.[20]

Watson wondered if the same mechanism might also account for phobias. Perhaps, reasoned Watson, phobic reactions were simply the result of an object or situation having being paired with a stimulus that made people behave as if they were afraid. To find out, Watson followed in Freud's footsteps and conducted a study on an unsuspecting infant. In 1919 Watson teamed up with a student named Rosalie Rayner, and together they experimented with an eleven-month-old boy they referred to as "Albert B." Watson speculated that if Albert could be made to behave as if he were afraid of something, he would quickly develop a Pavlovian response to that object and become phobic about it. Perhaps influenced by his laborious work with mazes, Watson decided to try to make Albert phobic about rats.

Before starting the study Watson needed to know that Albert didn't have a phobia of rats and so showed him a variety of rats and ratlike objects, including a rabbit, a monkey, and various hairy masks. The fearless Albert didn't flinch. Next, the intrepid researchers wanted to get Albert to associate the rat with something that made him behave as if he were afraid. Watson knew that infants jump out of their skin whenever they hear a very loud noise, and so went out and bought a large steel bar and a hammer.

Watson and Rayner then placed a white rat near Albert, and whenever their young subject went to touch the rat, they hit the steel bar with the hammer as hard as they could. Exactly as planned, the resulting bang made Albert start to cry. After several "rat—bang" trials, Watson stopped banging the bar and showed Albert the rat. In the same way that Pavlov's dogs came to salivate when they heard the bell, so the mere sight of the rat caused Albert to become extremely distressed. Watson had indeed created a phobia.

Two months later, Watson and Rayner visited Albert again and discovered that he was still terrified by the sight of the rat. Not only that, his fear had generalized to other similarly furry objects, including a dog, a seal-skin coat, and Watson wearing a Santa Claus mask.

Could the same idea explain Little Hans's fear of horses? When discussing the onset of Hans's phobia, his father described Hans as having been terrified by the sight of a carthorse collapsing in a local park and being especially frightened by the sound of its hooves clattering against the cobbled road. Hans's phobia was nothing to do with repressed sexual thought about his mother or concerns about masturbation. It was simply a Pavlovian response to a situation that had frightened him.

Little is known about what happened to Albert following the study, although Watson did joke that when he was older, a Freudian analyst would probably manage to persuade him that his fear of fur was due to his being told off when he was trying to play with his mother's pubic hair when he was three years old. In contrast, a great deal is known about the adult life of Freud's unsuspecting subject. "Little Hans" was actually the pseudonym for Herbert Graf. Graf went on to enjoy a highly successful career as an opera producer and is perhaps best remembered for his highly innovative adaptation of

Wagner's *The Ring of the Nibelung*, in which Brünnhilde's horse was replaced by a giraffe with an especially long neck.û*

The experiment with Albert had a huge impact on Watson's personal life. While he was running the study, Watson, a married man, had an affair with his co-experimenter Rosalie Rayner. When Watson's wife found out about the relationship, she sued for divorce, and when the president of John Hopkins University heard what was happening, he demanded Watson's resignation. Watson left academia and went to work for a major advertising company, using his behavioral insights to help sell deodorants, baby powder, and cigarettes. In one of his most successful campaigns, Watson introduced the idea of a coffee break to America as part of a campaign for Maxwell House.

Once psychologists understood how phobias developed, they were quickly able to understand more about the disorder and discovered that it was fairly easy to eradicate the problem.

Closer and Closer

One of the most effective procedures, systematic desensitization, was created by South African psychiatrist Joseph Wolpe (see the "Overcoming Phobias" box). During this technique, people are first trained how to relax. Next, they are asked to establish a "continuum of anxiety" that runs from a not very scary version of the phobia to a full-blown terrifying situation. So, for example, if a person was afraid of snakes, one end of the anxiety continuum might be opening a book and seeing a picture of a snake and the other end might involve meeting a real estate agent. At the start of the first session the person is encouraged to relax and then experience a situation (or, in some cases, imagine it) from the lower end of the anxiety continuum. Because the person is acting as if he or she is not afraid, the situation quickly ceases to become associated with any sense of anxiety, and the person moves on to the next situation on the continuum.

---

* I added the bit about the giraffe.

About 10 percent of the population suffers from some kind of phobia, and for about 1 percent, this disorder has a dramatic and debilitating effect.[21] They could be afraid of open spaces, humiliation, the sight of blood, and even the number thirteen. They often end up talking about their fears with a psychotherapist who attempts to uncover the deep-seated nature of the problem. They are wasting their time. From snakes to spiders and flying to fear of public speaking, there is a much quicker and more effective way forward: by changing their behavior one small step at a time, they will slowly change their mind forever.

Overcoming Phobias*

The techniques designed to help people overcome phobias often have three stages:

1. *Learning how to relax.* See the previous box, "Calming Down."
2. *Constructing an anxiety continuum.* People are encouraged to write down ten events that make them feel anxious and are associated with their phobia. They then grade the anxiety associated with each event by assigning it a number between 0 (very low) and 100 (very high). For example, if you have a fear of flying, your anxiety continuum might be: packing luggage, making reservations, driving to the airport, checking in, boarding the plane, taxiing, climbing to cruising

---

* The exercises described here are designed to provide general insight into the sorts of techniques that are used by psychologists. If you believe that you have a serious and problematic phobia, consult a professional.

altitude, moving around the cabin, turbulence, landing, crashing.

3. *The pairing procedure.* Finally, people are asked to go through their relaxation procedure and then experience the first item in the continuum for as long as they can (if it is not possible to actually experience the item, people are asked to imagine experiencing it). After they have stopped experiencing or imagining the situation, they are asked to rate how anxious they feel using the scale of 0 to 100. They then repeat this procedure until their anxiety rating for that event is below 10. When that happens, they move on to the next item on the continuum. Each session lasts around thirty minutes.

Hitting the Panic Button

About 5 percent of people have experienced a panic attack. The symptoms are obvious and unpleasant. Without warning, they start to experience chest pains and sweat, they hyperventilate, and they feel faint. During the episode, they might think that they are losing their mind or are about to die. The attacks usually last about ten minutes and then dissipate over the course of an hour or so.

Many doctors and psychoanalysts used to try to prevent these episodes by pumping people full of drugs or talking to them about their childhood memories. In fact, there is a simple explanation for them and a quick and effective solution.

In the previous chapter, I described the work of Stanley Schachter, who showed that experiencing an emotion is often a two-step process. First, an event or thought causes your body to jump into action. Perhaps you hear a gunshot, and suddenly your palms become sweaty, or maybe you catch the eye of an attractive stranger at a party and feel your heart miss a beat. Second, you look around and try to work out what caused your body to act in this way. If you were in the street when you heard the gunshot, you might feel anxious, whereas if you were walking past a shooting gallery at a fair, you might feel fine. Similarly, if you believe that the person at the party finds you attractive, you might experience a strong sense of joy, whereas if it turns out that this person was looking longingly at the person behind you, then you might feel a tad embarrassed.

In the 1990s, psychologist David Clark from the University of Oxford applied Schachter's thinking to panic disorders.[22] Clark thought that such attacks occurred when people misinterpreted their bodily sensations in a catastrophic way. According to this idea, people who are prone to panic attacks have a tendency to feel their heart race and palms sweat, and they assume the worst. Convinced that they are about to have a heart attack or die, they become even more stressed, which makes their heart beat even faster and their palms sweat even more. The process feeds on itself until they enter a state of extreme panic.

Clark believed that treating these attacks didn't need to involve taking drugs or discussing childhood memories. Instead it was a case of either preventing the sensations in the first place by teaching people how to relax or, even better, encouraging them to reinterpret their bodily sensations in a more productive way.

To discover if he was correct, Clark gathered together a group of patients prone to panic attacks and asked them to see themselves in a new way. The patients were told that they shouldn't panic when they felt their heart race or suddenly felt short of breath, but instead to see such sensations as evidence that their bodies felt slightly anxious. Some patients were worried that they might pass out during their panic attack, even though this had never actually happened. Clark eased their fears by explaining that this feeling was due to the blood traveling away from the brain and to the major muscles, and that the resulting increase in blood pressure meant that they were actually less likely to faint. Clark's procedure proved remarkably effective, with research showing that getting people to reinterpret their bodily sensations was more effective than either relaxation therapy or drugs.[23]

The same approach has been used to treat many of those who are overly anxious about taking exams, going to a job interview, speaking in public, or being hospitalized. In each instance, people have benefited from learning why their bodies sometimes get overly excited and how to reinterpret these feelings in a more positive way ("Exam nerves usually help focus attention," "A little extra adrenaline makes for a better interview or talk," "It is perfectly natural to be nervous before being hospitalized").[24]

An understanding of the way in which the body creates emotion has led to the development of quick and effective treatments for anger management, phobias, panic attacks, and some anxiety disorders. But can it also help with one of the most prevalent and problematic of psychological problems: depression?

The Guilt Trip

Ever been ashamed of something you have done? Worry not, because the As If principle can help to ease your guilt.

Simone Schnall from the University of Plymouth knew that unethical behavior often produces a look of disgust, but wondered whether forming a disgusted facial expression makes people think that an event is more unethical. To find out, Schnall and her colleagues armed themselves with several descriptions of morally questionable scenarios and a can of fart spray.[25]

Schnall's experiment took place on a busy main street close to a trash can. The researchers stopped passers-by, asked them to read short scenarios describing various types of questionable behavior (including items about first cousins marrying, a man knocking over a dog in his car and then eating the dog's body, and a man being sexually aroused by his pet kitten) and then rate how moral they found each situation. Beforehand the researchers had either applied a liberal dose of fart spray to the trash can or left it alone. The passers-by who were subjected to the fart spray made disgusted facial expressions and also judged the scenarios as significantly more immoral.

Encouraged by this work, other researchers had participants remember a time when they had acted in an immoral way, asked them to clean their hands with an antiseptic wipe, and then had them rate how guilty they felt and whether they would like to volunteer for charity work.[26] Washing their hands made people feel significantly less guilty.

So to get over your little moral indiscretion, go and clean your hands and let the As If principle wash away your sins. When you face even larger guilt trips, try taking a shower.

III. Dealing with Depression

According to the Bible, King Saul was not a happy man. Chosen to be king of Israel mainly because of his height ("he was higher than any of the people from his shoulders and upward"), Saul became embroiled in several wars and was frequently very moody ("an evil spirit from the Lord troubled him"). Eventually a young musician named David was summoned to the king's court, and his harp playing proved surprisingly effective at alleviating Saul's bad moods. A man of many talents, David then went on to slay Goliath and defeat the Philistines. This in turn made Saul jealous of the harp player's success and caused him to try to smite David with a javelin.

Liubov Ben-Noun from Ben-Gurion University of the Negev in Israel has carefully analyzed the biblical description of Saul in an attempt to diagnose his psychological difficulties using modern psychiatric criteria.[27] After ruling out substance-induced mood disorders (the Bible makes no mention of Saul taking drugs) and schizophrenia (Saul is one of the few people in the Bible not to hear voices), Ben-Noun concludes that Saul may have been suffering from major depression. In a later paper, Martijn Huisman from the University Medical Center Groningen in the Netherlands suggests that this disorder was probably brought on by work-related stress due in part to Saul's army of just three thousand men having to face Philistine forces consisting of "thirty thousand chariots, and six thousand horsemen, and people as the sand which is on the seashore in multitude."[28]

Modern epidemiology demonstrates that depression is not confined to the pages of the good book. It is estimated that in most

101

Western countries, one in twenty people will be suffering from depression at any one time, with some research suggesting that women are nearly twice as likely as men to struggle with the problem.

The symptoms include feelings of hopelessness, finding it difficult to get out of bed, avoiding contact with other people, a dramatic loss or gain in appetite, an inability to concentrate, and sleep problems. When trying to explain what it feels like to be depressed, some clinicians ask people to imagine "the anguish of grief with the sluggishness of jet lag."[29] Although everyone feels sad from time to time, in depressed individuals these symptoms continue for long periods and so have a more damaging effect on their lives. Sometimes depression appears to emerge as a response to a negative event, such as losing your job or the death of a loved one, and at other times, it just seems to emerge without any apparent cause. Although fierce arguments have been waged over the causes and treatment of depression, most physicians, psychologists, and psychotherapists originally believed that the solution to the problem involved changing what was happening inside people's heads.

During the 1940s, an American physician, Walter Freeman, believed that people became depressed because of a problem in the signals traveling between the front and center of the brain. Based on this reasoning, Freeman developed a strange medical procedure to cut the communication channels between these two parts of the brain. During this operation, the patient would first be rendered unconscious by an electric shock. Next, Freeman would insert the end of an ice pick like instrument called a leucotome through the patient's tear duct. After a few taps with a surgical hammer, the leucotome would move into the patient's frontal lobe, where Freeman would wiggle it about to destroy the offending parts of the brain. Freeman carried out over three thousand of these operations during his career and, ever the showman, would sometimes set up a production line of patients, once lobotomizing twenty-five women in one day.

His operation became known as a frontal lobotomy (a term derived from the Greek word *lobos,* meaning "lobes of the brain," and *tomos,* meaning "barbaric and ineffective hogwash"). Although some of Freeman's patients improved after their brief encounter with the leucotome, many suffered terrible side effects, including having to be retaught basic functions such as eating and drinking. Perhaps Freeman's highest-profile failure was President John F. Kennedy's sister, Rosemary. After experiencing dramatic mood swings and the occasional violent outburst, Rosemary was subjected to a frontal lobotomy when she was twenty-three years old. Rosemary was incapacitated by the operation and spent the remaining years of her life requiring full-time care because of severe learning difficulties and urinary incontinence.

After seeing the negative effects of destroying patients' frontal lobes, other physicians started to examine less traumatic ways of changing people's brains. By far the most popular procedure to emerge has been antidepressant drugs. Electrical impulses travel from one part of the brain to another via cells known as neurons. These neurons communicate with one other by releasing a chemical known as serotonin, and shortly after each communication, the serotonin is reabsorbed back into the neuron. In the 1960s, scientists discovered that high levels of serotonin in the brain were associated with a good mood, and so developed drugs that blocked the neuronal reabsorption of the serotonin in the hope that this would help cure depression. Although there is considerable controversy about the effectiveness of these drugs and their potential side effects, many researchers claim that they help reduce depression, and they are now one of the most popular forms of medication.

Reluctant to put pickaxes into people's brains or pump them full of drugs, psychologists have developed other ways of changing how people think.

The Allure of Attribution

Imagine that you take an exam and get a poor grade. How, in general, would you explain your failure? People come up with all sorts of replies to this question. Some say that perhaps they didn't study hard enough, others may say they just had an off day, or perhaps the questions they had expected and studied specifically for didn't come up. Psychologists think that you can tell a great deal about people from their answer and tend to rate replies in three ways.

First, did you blame yourself? If you thought that you failed the imaginary exam because you are not especially bright or didn't study properly, then you are blaming yourself for your failure. If you thought that you had an off day or got unlucky with the questions, you are blaming something outside yourself.

Second, there is the issue of permanence. Does your explanation suggest that you will do badly in future exams? Thinking that you are not very bright suggests that you will fail again in the future, whereas believing that you had an off day means that there is no reason to expect other tests to go badly.

Finally, what does your explanation say about what will happen in other areas of your life? If, for example, you think that you are not especially bright or are especially lazy, then you might expect to do badly at work or in the local trivia quiz at a bar. However, thinking that you failed the exam because you had an off day wouldn't have any implications for your career or other situations.

When bad things happen to depressed people, they tend to come up with explanations in which they take the blame, causing them to expect to fail again in the future and casting a dark cloud over every aspect of their lives. In contrast, those who are not depressed are far more likely to avoid personal shortcomings, expect the future to be bright, and prevent the failure from influencing other areas of their life.

When treating people with depression, psychotherapists often try to encourage their clients to recognize and change their explanations for the events that happen in their lives. This technique forms a central part of cognitive therapy (CT). Other aspects involve getting clients to identify additional types of problematic thinking, including "mind reading" (where they jump to conclusions about what other people are thinking), "catastrophizing" (where they transform into drama queens and make mountains out of molehills), and "fusion" (where they confuse their beliefs for facts).

Researchers have conducted several studies comparing CT with drug-based therapies and have discovered that they are equally effective at tackling depression.[30] Because of this, governments and health systems across the world have adopted the approach, encouraging millions of people to think about their thinking. After years of experimenting with different cures, all looked well. In fact, this was far from the full picture.

From Behavior to Memory

Let's conduct a small-scale experiment.

Stage One: Spend a few moments painting a happy expression on your face. Pull the sides of your mouth back toward your ears and hold them there. Sit or stand up straight, pull your shoulders back, and push your chest forward. Now look at the three words below, and for each word, think of an event in your life that is associated with the word. Next, get a piece of paper and jot down a couple of words from each memory in order to remind you of the experience.

tree   Words to remind you of your
memory_____

house Words to remind you of your
memory_____

cat   Words to remind you of your
memory_____

Stage Two: Spend a few moments forcing your face into a frown. Pull the sides of your mouth down, and hold them in that position. Next, if you are sitting down, slump forward.

If you are standing up, let your shoulders drop down. Now look at the three words below, and for each word, jot down an event from your life that is associated with the word.

boat   Words to remind you of your

           memory_____

car   Words to remind you of your

           memory_____

dog   Words to remind you of your

           memory_____

Stage Three: Look at the six memories that you have just described. Would you say that they were positive or negative?

This study was first conducted by researchers Simone Schnall and James Laird from Clark University.[31] They discovered that when people adopted a happy facial expression, they tended to remember more positive moments from their life, and when they looked sad, they tended to recall more negative ones.

Depressed people often tend to dwell on the aspects of their life that have not gone especially well, and this research suggests that their biased memory might be due in part to their behavior. So if you want to remember how great your life has been, smile and sit up straight, and let your brain do the rest.

Activating the Mind

The As If principle suggests that behavior causes emotion and helps explain why some people experience anger management issues, phobias, and panic attacks. Can the theory also help to explain depression? Could it be, for example, that depressives struggle to get out of bed in the morning not just because they feel down, but they also feel down because they are spending too much time in bed? A large body of research now suggests that this is indeed the case.

Much of the early work into the As If principle focused on facial expressions and emotion, and showed that smiling made people happy and frowning made them sad. Clinical psychologists have discovered a similar relationship between expressions and depression. In one study, for instance, Jessie Van Swearingen from the University of Pittsburgh recruited a group of patients suffering from a facial neuromuscular disorder and measured both the degree to which the patients were able to smile and their level of depression. As predicted by the As If principle, the less animated the patients' facial expressions were, the more likely they were to be depressed.[32] Similarly, dermatologist Eric Finzi has evaluated whether Botox injections can minimize some of the facial expressions associated with sadness and therefore help alleviate depression. In one small-scale pilot study, Finzi injected Botox into the frown lines of nine depressed women and then tracked their lives. The injections would have caused the women to frown less but not prevented them from forming other facial expressions. As a result, the researchers predicted that the procedure would help prevent the women from feeling sad and so help alleviate negative emotions. They were right: just two months after the injections, none of the nine women showed signs of depression.

Other work has taken a more behavioral approach and looked at the effect of dance on depression. Driven by the notion that dancing is incompatible with feeling down, Sabine Koch from the University of Heidelberg and her colleagues examined the impact of dance on depression.[33] Koch assembled a group of people suffering from depression and had them dance to upbeat music. Worried that any effects might be due to the music or moving around, Koch had other groups of participants listen to the same soundtrack or spend time on an exercise bike. All three groups felt better after their sessions, but those who had been dancing the night away showed the most impressive improvements.

Psychologist Peter Lewinsohn wondered whether it might be possible to help change the way in which depressives thought and felt by changing their behavior.[34] Depressive behavior is often about escape and avoidance. When some people encounter a negative life event, such as being laid off or the breakup of a relationship, they withdraw from the world in order to prevent more pain in the future. This withdrawal can take many forms, including spending large amounts of time in bed, avoiding their friends, comfort eating, excessive drinking, and drug taking. In addition, the person may also try to avoid thinking about future events by, for instance, ruminating on the past ("If only things had been different") or watching soap operas and quiz shows on television. Unfortunately, all of this has unintended and negative consequences. Lying in bed and overeating might make them put on weight and so feel ashamed of themselves. Excessive sleeping and television watching might encourage their partner to criticize them. And not contacting their friends is likely to decrease the chances of being invited out, thus increasing the feeling of isolation.

To help reverse this downward cycle, Lewinsohn created a simple technique, behavioral activation. There are several versions of the treatment, but most have two main phases.[35] In the initial part, people are encouraged to identify behaviors that are problematic and set some general goals (see the "Behavioral Activation: Phase One" box). This can involve their indicating which aspects of their behavior are symptomatic of depression and also identifying desired goals.

Behavioral Activation: Phase One*

In the first part of the process, psychologists use the following types of techniques to identify behaviors that are problematic and set general goals:

---

* The exercises described here are designed to provide a general insight into the sorts of techniques that are used by psychologists. If you believe that you are depressed, consult a professional.

1. Identify problematic behaviors.

On a sheet of paper mark down your answers to the following questionnaire.

Behavior	Yes/No
Are you avoiding seeing your friends and family?	Yes/No
Have you stopped taking part in activities that you enjoy, such as playing sports, going to the movies, or eating out?	Yes/No
Have you stopped taking care of yourself by not eating properly or looking after your personal hygiene?	Yes/No
Have you stopped trying to do well at school, college, or at work?	Yes/No
Do you tend to avoid thinking about the future by dwelling on your past?	Yes/No
Have you lost interest in your relationship with your children or partner?	Yes/No
Are you spending too much time watching television, playing computer games, or lying in bed?	Yes/No
Are you drinking heavily, comfort eating, or taking drugs?	Yes/No

Look at the list of your selected behaviors. Which of them would you like to change?

2. Identify desired goals.
3. Look at the following list of areas. Find one or two that you value and also one or two that you seem to be struggling with, and then answer the questions associated with the areas:

Your relationships: Would you like to be in a relationship or improve an existing relationship? Perhaps increase the number of friends you have or develop a better relationship with your parents or partner?

Work and education: Do you want to do well in college or improve your career? Perhaps run your own business, be promoted, or obtain a qualification or training?

Recreation: Do you want to get more enjoyment from your leisure time? What types of sports, interests, and hobbies would you liked to be more involved with?

Community: Would you like to do more for your community? Perhaps carry out some charity or volunteer work or engage in some form of activism?

Physical health: Would you like to be healthier? Perhaps lose some weight, exercise more, or eat a healthier diet?

In the second phase of the process, people are encouraged to engage in the type of activities that they have been avoiding and work toward desired goals (see the "Behavioral Activation: Phase Two" box). The emphasis is on behavior rather than on what is going on between their ears. Gone are the days of asking people how they feel; instead it is all about how they intend to change their behavior.

People are asked to create a list of specific activities that over time will help them change their behavior. For example, the general goal of spending more time with other people might mean having coffee with a friend once a week and going out to see a movie with work colleagues once or twice a month. Similarly, the general goal of gaining a new qualification might involve several specific activities, such as using the Internet to find out about possible courses and having a meeting with a boss to talk about taking time off from work to study. During this phase, worksheets are created to help motivate and monitor these behavioral changes.

Behavioral Activation: Phase Two*

1. Identify target behaviors.

Look back at the behaviors from the previous box that you want to avoid and your desired goals. For each one, make a list of specific activities that will help you avoid the former and achieve the latter.

Each of these specific activities should involve a small but real step toward your goal. So, for example, spending less time in bed might involve getting up at 9:00 a.m. every weekday and not going to bed before 11:00 p.m. Similarly, starting a new relationship might mean registering with an online dating agency, telling your friends that you are actively looking for a partner, and joining a book club.

All of these specific activities need to be measurable, realistic, and time specific. Writing "be happier" wouldn't be acceptable because it is difficult to measure and not time specific; "reading one new book every two weeks" would be much better.

A sample list of specific activities might include these:

- Wake up every day at 9:00 a.m. and get out of bed.
- Visit one museum or art gallery every week.
- Telephone my parents twice a week.
- Once a week, contact a friend and suggest that we meet for coffee.
- Each week, write 500 words of my novel.

---

* The exercises described here are designed to provide a general insight into the sorts of techniques that are used by psychologists. If you believe that you are depressed, consult a professional.

2. Create a plan.

Use this chart to create a plan for each day of the week, noting which of the specific activities that you intend to achieve and when you will achieve them.

MONDAY: Date:			
Time	Planned activity	Actual activity	Rating of how successful this was between 1 (not very successful) and 10 (very successful)
9:00 a.m.	Wake up and get out of bed.		
10:00 a.m.			
11:00 a.m.	Telephone my parents.		
12:00 a.m.	Produce 100 words toward my novel.		

At the end of the week, review the tables and identify which targets you did, and didn't, achieve. Carry the targets that you didn't achieve over to the next week. You might find the following tips helpful:

• Don't try to change all aspects of your behavior at once. Instead, start with baby steps and build up gradually.
• Don't let your thoughts get in the way. If you feel yourself think-

ing about failure or feeling bad about yourself, just accept these thoughts and move on.

- Everyone fails once in a while, so if you don't manage to achieve your goals, don't worry. Make up another activity sheet and try again.
- Moving out of your comfort zone may feel difficult at first, and it might feel better to fall back into your normal routine of avoidance by telling yourself, "I'll do it when I feel better," or "I will wait for exactly the right situation."

Don't fall into these traps: try to change your behavior no matter how you feel or what you are thinking.

According to the As If principle, the behavior activation technique should be effective, but is it?

In 2006, Sona Dimidjian from the University of Washington and her colleagues conducted a remarkable study.[36] Dimidjian recruited two hundred outpatients suffering from major depression and randomly assigned them to one of four groups. Group 1 was given a course of paroxetine, a popular antidepressant drug; group 2 was given an inert placebo pill; group 3 attended cognitive therapy; and group 4 went through behavioral activation.

The research team then tracked the patients over two months to discover which of the treatments worked best. The results revealed that for the most severely depressed patients, behavioral activation was significantly more effective than cognitive therapy. Perhaps most important, the study showed that behavioral activation was as effective as taking paroxetine.

Over the years, a large number of studies have demonstrated this effect time and again.[37] When it comes to alleviating depression, trying to change what goes on in people's heads with drugs and cognitive therapy can be tricky. In contrast, changing their behavior has far fewer side effects and is just as effective.

The As If principle is not just about producing happiness and love. It can also help eliminate pain and suffering and, in doing so, help millions of people live better and more productive lives.

Twenty Pieces: Part One

Before you turn to the next chapter, complete this exercise.

First, take a letter-size piece of paper and rip it into twenty pieces. Each of the pieces can be any size and shape you like. It is a fairly boring task and will take just a few minutes. You can do it now or leave it until later.

I will remind you again before the end of the next chapter.

Chapter 4

Willpower

Where we learn why rewards punish, and discover how to motivate others, beat procrastination, stop smoking, and lose weight

"I prayed for twenty years but received no answer until I prayed with my legs."

—Frederick Douglass

I. How Rewards Fail and What to Do about It
II. Why Small Changes Have a Big Impact
III. Losing Weight without Trying

I. How Rewards Fail and What to Do about It

Psychologists have long tried to unpack the mystery of motivation. Why is it that some people are self-controlled and driven, while others find it difficult to drag themselves out of bed in the morning? During the 1960s, much of the work that researchers conducted used pigeons placed in specially constructed cages and then carefully observed their bird-brained participants. The cages contained a button and a light, and the researchers attempted to train the pigeons to peck at the button whenever the light was illuminated. Extensive experimentation soon revealed that the pigeons learned much more quickly when they were rewarded with small pellets of food. Assuming that people were much like large, featherless pigeons, many researchers believed that the same type of reward system could be used to motivate humans. The idea was quickly embraced by organizations and governments across the globe: prisoners were now allotted special privileges for good behavior, schoolchildren were offered candy when they read books, and employees were awarded bonuses when they were especially productive. Unfortunately, it soon became clear that laboratory-based pigeon research could not be generalized to humans living in the real world. Some of these reward systems either had no long-term effect or actually deterred the very behavior that they were designed to encourage.

In his book *Punished by Rewards,* Alfie Kohn catalogs the wealth of evidence demonstrating the downside of incentives.[1] In one study, for example, researchers tracked more than a thousand people as they attempted to give up smoking.[2] The investigators randomly split the smokers into two groups and asked everyone to take part in an eight-week course designed to help them quit. The participants in one of the groups were given various incentives to take part in the antismoking program, including a free ceramic mug and an opportunity to win an all-expenses-paid trip to Hawaii. The smokers in the other group acted as a control and so were not given any incentives. Initially the rewards worked well: the participants who had been given the ceramic mug and dreams of sun-drenched beaches were especially enthusiastic about the program. However, when the researchers revisited the participants three months after the start of the study, they discovered that the same percentage of people in both the control group and incentives group had abstained from smoking. After a year, more people in the incentivized group than in the control group had relapsed.

In another investigation psychologist E. Scott Geller from the Virginia Polytechnic Institute reviewed twenty-eight studies that had encouraged people to use seat belts.[3] After looking at the data from almost 250,000 people across a six-year period, Geller concluded that rewarding people with cash or gifts for buckling up was one of the least effective ways of encouraging long-term use of seat belts. Similar large-scale reviews of reward-based programs designed to encourage schoolchildren to read also revealed no long-term benefits.[4]

Then there is the work on rewards for creativity. Offer artists big money. and you might think they will soon get their creative juices flowing. But when Teresa Amabile from Brandeis University in Massachusetts asked a panel of professional artists to judge the artistic merits of both commissioned and noncommissioned creative work (without knowing which was which), she discovered that the noncommissioned projects were awarded higher ratings than the commissioned ones.[5]

Worried that this might not be due to the negative influence of rewards, but rather artists having their style cramped by the demands of financial backers, Amabile carried out a more controlled investigation.[6] She recruited a group of budding writers and asked them to write a haiku-style poem using the word *snow* as the first and last line. The participants were then split into two groups. One group was asked to think about all of the riches that would flow from being a great writer, while the other group was asked to dwell on the intrinsic pleasure that they derived from their work. Finally, everyone was asked to produce a second poem around the concept of laughter.

Amabile then assembled a panel of twelve poets, gave them the haiku poems about snow and laughter, and asked them to rate how creative the poems were. Both groups displayed the same amount of creativity when writing about snow. However, the group that was made to think about the rewards and riches that might flow from their writing displayed significantly less creativity when writing about laughter. Even thinking about rewards had had a detrimental effect.

Many psychologists were stunned by the results. Why should the reward systems that worked so well in the laboratory frequently fail in everyday life?

Why Rewards Punish

Spend a significant amount of time with any social psychologist, and sooner or later he or she will tell you the story about the wise old man and the abusive teenagers.

According to the anecdote, there was once a wise old man who lived in a rough neighborhood. One day a group of surly teenagers decided to give the old man a hard time. Each day they would walk past the old man's house and shout abuse at him. Many old men would have decided that the best way forward was to shout back at the teenagers, call the police, or hope that the group would eventually grow bored of their mean-spirited ways. However, the wise old man had a sound understanding of psychology and so came up with a completely different, and altogether more cunning, plan.

He sat outside his house waiting for the teenagers to come along. When the group showed up, the old man immediately handed a five-pound note to each of the teenagers and explained that he was happy to pay them to shout abuse. Bemused, the teenagers took the money and came out with their usual rude remarks. The old man repeated his actions each day for a week.

The following week was slightly different. When the teenagers came along, the old man explained that he had had a tough week financially, and so could pay each of them only a pound. Unperturbed, the teenagers continued to take his money and persisted with their childish chants.

At the start of the third week, everything changed again. When the group arrived, the old man explained that it had been another tough week, and now he could afford to pay the teenagers only twenty pence each. Insulted at the small amount of money on offer, the teenagers refused to shout their abuse.

The story is almost certainly apocryphal, but it reflects a fundamental truth about why we do what we do. To fully understand the wisdom behind the old man's actions, we need to travel back to the 1970s and discover what happened when a group of people were paid to solve a wooden puzzle.

Psychiatrist Edward Deci was a big fan of a commercially available puzzle called Soma. The puzzle involved giving people several oddly shaped wooden blocks and asking them to arrange the blocks into specified shapes. Deci wondered if the puzzle could be used to discover whether the As If principle influences motivation.[7]

Deci invited volunteers into his laboratory and asked them to play with the puzzle for thirty minutes. Before starting, some of the volunteers were told that if they could solve the puzzle. they would be given a financial reward. The others weren't offered any incentive.

After thirty minutes, Deci told all the participants that their quality time with Soma was up. He then explained that he had left the paperwork needed for the next part of the experiment in his office and so had to leave the laboratory to fetch it. As is so often the case with social psychology experiments, this "I have to leave the laboratory now" ploy was a cover story. The important part of the experiment was just about to take place.

Deci left each participant alone for ten minutes. During this time, they were free to continue to play Soma, or read the magazines that had been strategically placed on a nearby table, or indeed neither. All of the time, Deci was secretly observing their behavior.

Conventional pigeon-based reward theory predicts that those who had been paid to play Soma would have found the puzzle especially enjoyable, and so be more likely to continue with it when Deci left the laboratory. In contrast, the As If principle makes a quite different set of predictions.

According to the As If principle, the participants who were offered the financial reward to play with the puzzle would have unconsciously thought, "People offer me money only when they want me to do something that I don't want to do. I was offered money to play with the puzzle, so it can't be much fun." Using the same logic, those who weren't offered any financial incentive would have unconsciously thought, "People offer me money only when they want me to do something that I don't want to do. I wasn't offered any money to play with the puzzle, so it must be fun." Seen in this way, those who had been offered the reward had been made to behave as if they didn't actually enjoy playing with the puzzle, whereas those who hadn't been offered any financial incentive were behaving as if playing with the puzzle was fun. According to the As If principle, Deci's payments would have turned play into hard work, and so the participants who were financially rewarded would be significantly more likely to leave the puzzle alone when he left the room.

Deci's results provided overwhelming support for the power of acting As If. Regardless of participants' success at completing the puzzle, those who had not been previously offered the financial reward were far more likely to play with the puzzle when left to their own devices.

Other researchers quickly conducted several similar experiments to discover whether this intriguing finding was genuine. In perhaps the best known of these studies, Stanford psychologist Mark Lepper and colleagues visited several schools and asked schoolchildren to draw some pictures.[8] Before being allowed access to the crayons and paper, Lepper told one group of children that they would receive a "good player" medal for drawing. A second group of children was not given the promise of any reward. According to the As If principle, the children who were offered the medal would have unconsciously thought, "I am offered rewards only when adults want me to do something that I don't enjoy. I am being offered a gold medal for drawing; therefore, I must not like drawing." Similarly, the children in the other group would have thought, "I am offered rewards only when adults want me to do something that I don't enjoy. I am not being offered any reward for drawing; therefore, I must like drawing."

A few weeks later, Lepper and the teams returned to the school, handed out the drawing materials again, and measured how much the children played with them. The children who had received the medals a few weeks before spent significantly less time drawing than their classmates did.

The message from the studies is clear: rewarding the behavior of schoolchildren, smokers, and drivers encourages them to behave as if they don't really want to read books, stop smoking, or buckle up. As a result, the moment the rewards are removed, the desired behavior runs the risk of grinding to a sudden halt or, worse, becomes even less frequent than before incentives were introduced. In the short term, reward systems can be effective. However, over the long haul, most organizations struggle to maintain a continuous supply of special privileges, candy, gifts, and bonuses, and the moment the rewards stop, people's motivation often vanishes into thin air.

The Man with X-Ray Eyes

Having established that the As If principle plays a key role in motivation, researchers began to explore other ways in which the effect could be used to encourage people to spring into action.

In the workplace, some business gurus argued for the importance of reconfiguring jobs to make them more intrinsically enjoyable by giving employees a greater sense of autonomy, purpose, and fun.

When it came to people's personal lives, some psychologists turned their attention to role play. Consider, for example, the work of Leon Mann from Harvard University and his groundbreaking study into giving up smoking.

Mann invited twenty-six very heavy smokers to his laboratory and randomly split them into two groups.[9] Those in one group were asked to behave as if they were going to give up smoking by role-playing someone who had been diagnosed with cancer. To make the pretense as realistic as possible, Mann created a fake doctor's office at the university. When the participants entered the room, they saw a variety of medical equipment and an actor wearing a white coat. The actor played the role of a doctor and brought out the participant's alleged X-rays. It wasn't good news: according to a set of fictitious medical records, the participant had lung cancer. The participant was asked to respond by discussing how he or she now intended to give up smoking. In contrast, the participants in the control group were shown the same highly emotional information about having lung cancer but were not asked to alter their behavior by taking part in any form of role play.

The results were remarkable. At the start of the study, all of the participants were smoking around twenty-five cigarettes a day. Immediately after the experiment, those in the control group cut down by an average of five cigarettes per day, whereas those who had been involved in the role play showed an average reduction of ten cigarettes per day. Acting as if they were going to reduce the amount they smoked caused a significant change in the participants' actual behavior. The researchers then tracked the participants over the next few years and discovered that the effect was not short-lived.[10] Two years after taking part in the study, those who had been involved in the role playing were still smoking significantly less than those in the control group.

When they weren't injecting meaning in the workplace or role play in people's personal lives, psychologists were busy discovering that a little often goes a surprisingly long way.

II. Why Small Changes Have a Big Impact

Let's imagine that you are at home and hear a knock on the door. You peek through your curtains and see a young man standing on your doorstep. He looks harmless enough, and so you decide to answer the door. The man explains that he is a volunteer working for an organization that supports research into cancer and wonders whether you would make a donation. You think about it and then decide that it is better to give than take, and so you give a small amount of money to him.

This may seem like an innocent encounter. In reality, you may have just taken part in a psychology experiment. This type of "would you mind contributing to charity" study was first carried out by Patricia Pliner from the University of Toronto and demonstrated how the As If principle could be used to get people to spring into action.[11]

Pliner's results showed that 46 percent of residents were prepared to write a check to the cancer charity. In the next stage of Pliner's experiment, the researchers had the volunteers approach a second set of houses and ask the residents to wear a lapel pin to help publicize their cause. The pin was small, and almost all of the residents agreed. Two weeks later, the volunteers returned to these pin-wearing residents and asked for a financial donation. Amazingly, over 90 percent of the residents agreed to give to the charity.

Known as the foot-in-the-door technique, this approach works because the small initial request caused the residents to behave as if they were the type of people who give to charity. This encourages them to believe that they are altruistic individuals, and so motivates them to agree to the much larger request. More than forty years of research have shown that the technique works in many different situations.[12]

Some of the most interesting and practical work has been conducted by French researcher Nicolas Guéguen. In one study, Guéguen traveled to Brittany and randomly split some of the residents into two groups.[13] Guéguen then telephoned the residents in one of the groups, pretended that he was from a local energy firm, and asked them to take part in a short telephone survey about energy conservation. A few days later, Guéguen sent out a letter to all of the residents in his experiment. The letter came from the mayor of the town and asked people to participate in an energy conservation program. More than 50 percent of those who had been asked to take part in the telephone survey agreed to participate compared with just 20 percent of those who hadn't been previously contacted.

In another study, Guéguen emailed more than a thousand people and asked them to visit a website that had been set up to support children who had been the victims of war.[14] When half of the people visited the website, they saw a message inviting them to click another link if they wanted to make a contribution to the charity. In contrast, when the others came to the website, they were asked to sign a petition against land mines and then asked to click a link if they wished to make a financial contribution to the charity. Only 3 percent of those who weren't asked to sign the petition clicked the donation link versus nearly 14 percent of those who had signed the petition.

Finally, Guéguen has also used this foot-in-the-door phenomenon to help Cupid find his target.[15] Venturing out onto the streets, Guéguen arranged for experimenters to approach more than three hundred young women and ask them out for a drink. Sometimes the experimenter asked for directions, or a light for a cigarette, before popping the question. Other times they took a more direct approach and asked the woman about the possibility of a drink straightaway. This small change made a big difference, with 60 percent of those who had been asked directions first saying yes to the drink versus just 20 percent of those who had been asked directly. In each instance, people had seen themselves acting as if they were pro-energy conservation, antiwar, or up for a drink, and so became motivated to act in a way that was consistent with their newly found identity.

Salespeople often turn to this powerful principle. Behavioral expert Robert Cialdini refers to it as the low-ball technique. It involves a set of procedures designed to get people to behave as if they are interested in a certain product or service.[16] For example, a car showroom may advertise a vehicle at a very reasonable price in order to lure potential customers into the showroom. It is only when they have inquired about a car, and therefore behaved as if they are interested in buying it, that the salesman explains about the various extra add-ons that will increase the price of the vehicle. Similarly, hotels may place online advertisements offering rooms at a low price. Once a potential customer has clicked on the ad, and so behaved as if he or she is going to make a reservation, the traveler discovers that the bargain rooms have all been sold but some higher-priced rooms are still available.

Twenty Pieces: Part Two

Just before the start of this chapter, I asked you to complete part one of the Twenty Pieces exercise. This exercise is all about procrastination. I pointed out that it was a fairly boring task, and that you didn't have to do the tearing there and then. Did you complete the exercise? If so, you are probably the type of person who doesn't have much of a problem motivating yourself when the going gets tough. However, if you decided to leave the tearing until later, then you are more likely to suffer from procrastination. The ability to procrastinate often stops people from doing well in many aspects of their lives and makes them feel tired and lacking

in control (or, as William James put it, "Nothing is so fa-
tiguing as the eternal hanging on of an uncompleted task").

If you do fall into the latter camp, then worry not: the
As If principle can help. Simply go back to the exercise, pick
up a piece of letter-size paper (no need to rip it into twenty
pieces just now), and then read the following paragraph.

How do you feel about completing the rest of the exercise now? According to research, at this moment you will
be experiencing a strange urge to finish the job by ripping
the page into twenty pieces. By working on an activity for
"just a few minutes" (that is, behaving as if you are a highly
motivated person), you change the way you see yourself and
make it far more likely that you will complete whatever it is
you have to do.[17]

Whenever you have a mountain to climb, persuade
yourself to spend just a few minutes taking those all-
important first few steps.

The principle can also cause people to radically change their behavior for the worse. In the early 1970s, for instance, the military
junta in Greece wanted to train ordinary soldiers to become sadistic
torturers.[18] Using the foot-in-the-door technique, they slowly persuaded these soldiers to abuse prisoners. At the start of the process,
the soldiers were asked to stand outside cells while prisoners were tortured. In the next stage of the process, they were invited to come into
the cells and witness the abuse. Then the soldiers were asked to provide a small amount of assistance inside the cells, perhaps holding the
prisoner down as he was beaten. Finally, they were asked to carry out
the beatings themselves, and so became the next generation of torturers as the newest recruits stood outside the cells. Slowly but surely, the
foot-in-the-door technique motivated the soldiers to carry out actions
that they would have originally considered completely unacceptable.

On a more positive note, some of the most recent research into the foot-in-the-door technique has explored whether the smallest of commitments can help make the world a better place.

Americans generate more than 150 million tons of trash every year, enough to fill the New Orleans Superdome twice every day.[19] Psychologist Shawn Burn from the California Polytechnic State University decided to discover whether the foot-in-the-door technique might be able to promote recycling.[20]

Burn's experiment involved five separate areas in Claremont, a small, affluent college town in eastern Los Angeles County. Just before the start of the experiment, Burn and his colleagues secretly observed the recycling activity of the residents, identified about two hundred households that didn't recycle, and set out to see if they could change the residents' behavior.

Burn started off by enrolling the help of a local troop of Boy Scouts and spent three weeks training them for the study. First, he had the scouts rehearse reading aloud a carefully created message stressing the need for recycling. Next, the investigators played the role of Claremont residents and asked the scouts to knock on their imaginary door and deliver the well-rehearsed message. When the investigators were convinced that their highly trained scouts were ready for the job, they were let loose on the public.

The scouts were put into groups of three. The investigators took each group to Claremont and asked the groups to knock on the door of an unsuspecting participant. When the door opened, the scouts would launch into their carefully prepared speech about the importance of recycling. A few moments later, they would hand the resident a pledge card and a sticker. The pledge card simply said, "I, _____, pledge support for Claremont's Recycling Program. I will help win the war on waste!" The sticker was equally straightforward and carried the message, "I Recycle to Win the War on Waste."

For the next six weeks, the research team again took to the streets and secretly observed the recycling behavior of the residents. The results were remarkable. Those who had not been visited showed only a 3 percent increase in recycling. In contrast, asking people to sign a pledge card and put a sticker up in their house resulted in a 20 percent shift. Just spending a few moments behaving as if they intended to recycle had a dramatic impact on their subsequent motivation to go green.

Change4Life

In 2011, I teamed up with the British government to help promote a campaign that used the foot-in-the-door technique to encourage the public to lead healthier lives. The work, part of a national campaign known as Change4Life, sought to encourage people to make small alterations to their diet and level of exercise in the hope that these would then lead to more significant changes.

In part of the work, we asked people to change their behavior whenever they encountered any of the following ten triggers. Try incorporating them into your life, and see if they act as a catalyst for more significant changes.

Trigger	Instead . . .
When you find yourself reaching for some sweets, chocolate, or potato chips . . .	. . . stop, and instead go for some fresh fruit (such as grapes, a banana, or raisins), rice cakes, or unsalted nuts.
When you are thinking about frying something . . .	. . . grill it instead. Try grilling your bacon and sausages, and scrambling or poaching your eggs.

When you are just about to order a large glass of white wine . . .	. . . change it for a spritzer made with a smaller amount of wine and sparkling water.
When you are out and about and just about to take the elevator or escalator . . .	. . . look around for some stairs and, if possible, use them instead.
If you are using public transport . . .	. . . see whether you can get off one stop too soon and walk the rest of the way.
When you are cooking for you or your family and are about to serve up the food on a full-size plate . . .	. . . swap the plate for a smaller one; it will encourage you to serve more modest portions.
When you're just about to put sugar into your tea or coffee . . .	. . . put in only half the amount of sugar.
When you are in a store and about to buy some white bread or white rice . . .	. . . boost the amount of fiber in your diet by choosing whole-wheat bread and brown rice.
When you are about to order a sugary soda . . .	. . . change your mind and go for sparkling water, milk, or pure fruit juice.
Instead of ordering a large main course . . .	. . . go for a smaller portion with a side order of salad or vegetables.

III. Losing Weight without Trying

Levels of obesity are rising across the world. In the 1980s 15 percent of Americans were considered obese. By 2003 this had risen to around 34 percent, with a staggering 17 percent of American children and adolescents classified as overweight. Carrying too much weight increases the likelihood of various health problems, especially heart disease, type 2 diabetes, and certain forms of cancer. It therefore isn't surprising that millions of people make a concerted attempt to lose weight at some point in their lives. Ironically, the chances that they will be successful are slim.

Many people are attracted to the quick-and-easy results promised by proponents of very low-calorie diets. Such diets are usually based around low-calorie, nutritionally complete liquid meals. The short-term impact of such diets is impressive: several research studies suggest that around half of those using these products quickly lose around 80 percent of their excess weight. However, when the researchers followed the dieters for the next few years, a very different picture emerged: after about three years, most of the participants had returned to their prediet weight, and after five years only three percent remained slim.[21] This depressing pattern of results is not limited to very low-calorie diets. After examining the results from hundreds of studies involving many different types of diets, one reviewer noted, "It is only the rate of weight regain, not the fact of weight regain, that appears open to debate."[22]

Pull Me—Push You

If you are sitting at a table, try this quick two-part exercise. First, close the book, put it down on the table, and push it away from you. Second, pull the book toward you, pick it up, and give it a little hug and a kiss (if you are doing this in a bookshop or another public place, you might want to smile at those around you in a "It's okay, I'm not dangerous" kind of a way).

How did you feel about the book after each part of the exercise? Research suggests that pushing an object away from you (behaving as if you don't like it) makes you dislike the object, whereas pulling it toward you (behaving as if you like it) makes you feel far more positively about it.[23] As far as I am aware, no previous research has examined the impact of hugging and kissing an object, but I am assuming that that makes you feel especially attached to the book.

Next time you are confronted with a plate of high-calorie snacks or chocolate chip cookies, simply push the plate away from you and feel the temptation fade. Similarly, if you are in the sales game and want to make a client feel more positive about a product, place it on a table in front of this client and encourage her to slide it toward herself.

Other work that has tried to promote weight loss through physical exercise has faced similar problems. In 2008 researcher Larry Webber from Tulane University and his colleagues reported the results from a large-scale study examining whether it was possible to encourage physical exercise in middle school children.[24] The two-year experiment encompassed thirty-six schools and thousands of children in the United States.

In half of the schools, the researchers did everything they could to encourage exercise and weight loss. Each week, the children were told about the importance of physical activity and asked to carry out a specified amount of exercise. The researchers even encouraged the schools to team up with local health clubs and recreation centers and stage special dance classes, gym sessions, and basketball matches. In contrast, the children in the other schools acted as a control and so weren't subjected to such opportunities and encouragement.

To discover the effect of the program, the researchers fit all children with an accelerometer to measure their movements and also tracked their body mass index. What impact did the extensive program of encouragement have? Almost zilch. The children who were given the opportunity to play more sports and encouraged to exercise ended up moving around very slightly more than those in the control group. Perhaps more important, there was no difference in the average body mass index between the two groups.

Why? The study was based on the notion that changing minds causes people to change their behavior. According to this approach, all you have to do is tell people about the importance of a healthy diet and regular exercise, and they will immediately jump into action. However, this approach has been shown to be flawed. Instead, an understanding of the As If principle provides a far more effective technique for long-term weight loss.

Eating with Your Eyes

Chapter 2 looked at the pioneering work of psychologist Stanley Schachter that revealed the surprising relationship between the As If principle and attraction. In the 1960s Schachter also came up with a similarly daring hypothesis to explain why some people are obese.

According to Schachter, people start eating on the basis of two very different types of signals.[25] The first set of signals comes from within your body. For example, after you have just consumed a large meal, your stomach might send you an "Okay, I couldn't even manage a wafer-thin mint" message, and so you know that it is time to stop eating. Or maybe you feel your tummy rumble, experience a sudden drop in blood sugar, and know that you should head to the kitchen. Conceptually, eating because you feel hungry is like feeling happy because you are smiling. In both instances, you decide how you are feeling on the basis of what your body is telling you.

Alternatively, your decision to eat might be influenced by signals from your surroundings. You might, for instance, see a great-looking cake in a bakery window and decide that it has your name written all over it. Or you might glance at your watch, see that it is time for a snack, and head for the kitchen. In these instances, you are ignoring the signals from your body and instead deciding how you should feel on the basis of what is happening around you.

Although everyone is influenced by both sets of signals, Schachter speculated that some people are more likely to listen to their body (Schachter referred to them as "internals"), while others are more likely to be affected by their surroundings ("externals"). He also hypothesized that when provisions are scarce, neither group will become obese, as internals will eat just when they are hungry and externals will munch away on the rare occasions that they have access to food.

So far, so good. However, in most developed countries, the supermarket aisles are jammed full of junk food, fast food chains encourage people to supersize, and movies offer huge buckets of popcorn with calorie-filled "butter." According to Schachter's theory, this excess of food will not present a problem to internals: they will continue to listen to their bodies and munch away when they are hungry. In contrast, the externals now have a problem. To them, each of the food mountains that they encounter on a daily basis screams out "eat me." Unless these people can exercise an extraordinary amount of self-control, they will soon find themselves consuming everything in sight. As a result, Schachter predicted that in most of the developed world, internals will tend to be on the slim side while externals will tend to be overweight.

It is an elegant and clever theory, but is it correct? To help find out, Richard Nisbett from Yale University conducted an ingenious experiment.[26] He recruited a mixture of slim and overweight people and invited them to his laboratory one at a time. All of the participants were asked to arrive in the early afternoon and not to eat after nine o'clock in the morning. After taking part in a boring experiment ("Can you count back from a thousand in threes?"), each of the participants was rewarded with some sandwiches. In reality, the tedious experiment was completely irrelevant; the core of Nisbett's experiment was to secretly observe the participants after they had been given the sandwiches (thus scientifically proving that there really is no such thing as a free lunch). Each participant was presented with a plate containing either one or three tasty roast beef sandwiches and told that they were free to help themselves to more sandwiches in a nearby refrigerator.

Schachter's theory predicted that the slim participants would be internals, and so the amount of food they consumed would not be related to the number of sandwiches on the plate. If they felt hungry, they would start to consume the sandwiches, and they would stop eating the moment they felt that their stomachs were full. In contrast, the overweight participants would tend to be externals who were driven by their eyes, and so would eat far more when they were presented with the three sandwiches. Obviously it was possible that the overweight participants felt hungrier than their slimmer counterparts. To allow for this, the experimenters made an intriguing prediction: following the out-of-sight, out-of-mind maxim, they predicted that the overweight participants wouldn't be any more likely than the slim subjects to raid the refrigerator.

What happened? When both groups of participants were presented with one sandwich, they ate the same amount. However, when they were presented with three sandwiches, the overweight participants quickly gobbled down far more than their slim counterparts. Not only that, both the slim and overweight participants were equally unlikely to raid the refrigerator.

In another cleverly designed study, Ronald Goldman and his colleagues from Columbia University made good use of Yom Kippur.[27] Yom Kippur is one of the most sacred of the Jewish holy days, with those adhering to the Jewish faith undergoing twenty-four hours without food or water. Goldman knew that modern Jewish people vary in the degree to which they observe this tradition, with some spending almost all of Yom Kippur in the synagogue and others going there for only hour or so. Goldman speculated that those who spent lots of time in the synagogue would not be constantly reminded about food. (In his paper describing the work, Goldman noted that the main food-related reference during the religious rituals at Yom Kippur would be a "passing mention of a scapegoat.")

On the basis of Schachter's theory, Goldman speculated that slim people would be relying on their bodies to tell them whether they were hungry, and so feel the same amount of food-deprived discomfort regardless of the amount of time that they spent at the synagogue. In contrast, overweight people would be relying on their surroundings to tell them whether they should eat and so feel significantly more comfortable at the synagogue. To find out if this was the case, Goldman sent around a questionnaire to his Jewish students asking them for their height, weight, how many hours they spent at a synagogue during Yom Kippur, and how difficult they found fasting. The results from the slim students revealed no relationship between the number of hours they spent at the synagogue and how difficult they found the fasting. In contrast, the longer the overweight students spent away from the synagogue during Yom Kippur, the more difficult they found fasting—exactly as Schachter had predicted.

Schachter's theory has huge implications for restaurants wanting to sell more food and people wanting to diet. From a restaurateur's point of view, persuading customers to become less self-conscious, and so ignore the signals from their stomach, is good for business.[28] For example, low lighting and soft music help focus people's attention away from themselves and also encourage diners to eat more. Similarly, tempting externals with photographs of food, or indeed the real thing, is also good for business. Research shows that placing alluring images of delicious-looking dishes on a menu, or wheeling in the dessert trolley at the end of a meal, will tempt even the most self-controlled of externals. In one study, researchers asked the staff at a French restaurant to classify their diners as either "chubby" or "normal" weight.[29] At the end of the meal, a server approached tables with a pie in hand and asked who would like dessert. Both the normal and overweight diners were equally likely to order dessert, but the overweight people showed a strong tendency to go for the pie that was sitting right in front of their eyes.

If you do want to lose weight, Schachter's theory can help. Try to get in touch with your inner internal by focusing more attention on what your body is trying to tell you. Just before you order that cake, for example, ask yourself, "Am I really hungry?" Similarly, reduce the tempting sights that are leading you astray by keeping unhealthy food out of sight and staying away from supermarket aisles packed with snacks and cookies. Also, try to avoid situations that involve eating food and distracting your attention away from yourself. Don't watch television, listen to music, or even read while you are eating. Instead, focus your attention on the food itself and slowly chew each mouthful. If that fails, try to make mealtimes as self-conscious as possible by eating in front of a mirror,[30] replacing your knife and fork with chopsticks (or vice versa), or using your nondominant hand.[31]

Schachter's simple theory links the As If principle with eating. Slim people are basing their decision about whether to eat on the signals from their bodies. In the same way that people feel happy when their face forms a smile, so they eat when their stomach tells them that they are hungry. In contrast, people who are overweight tend not to base their decisions about eating on the signals from their body and are instead more influenced by external signals. By getting them to refocus on behaving more in line with the As If principle, they can quickly shed any unwanted pounds. Dieting does not have to involve struggling in vain to overcome temptation. Instead, it is a question of listening to what your body is telling you.

Monitoring Yourself

Take a look at your computer monitor. Is the center of the monitor above, below, or level with your eye line? According to research exploring the As If principle and motivation, the position of your monitor could be having a significant effect on your productivity.

In the 1980s, John Riskind from Texas A&M University decided to examine the effect of posture on persistence.[32] He placed participants in one of two positions. Half of the participants were placed in a slumped position, such that their backs were stooped and hunched over and heads dropped down. In contrast, the other participants were made to sit upright, with their shoulders pulled back and heads held high. After sitting stooped or upright for about three minutes, participants were sent to another room and asked to try to solve several geometric puzzles that involved tracing over a diagram without lifting their pencil off the page. In fact, many of the puzzles were impossible to solve; Riskind was interested only in how long the participants kept going in the face of failure. Describing his findings in a paper entitled "They Stoop to Conquer," Riskind noted how the participants who had previously been sitting up straight persevered for almost twice as long as the slouchers.

More recently, other psychologists built on this work by asking participants to sit at a computer and work on a complex problem.[33] Some of the time the computer monitor was placed in a low position, causing participants to slump down. Other times it was placed slightly above their eye line, causing participants to sit up straight. Once again, those with their head held high persevered for longer.

To ensure maximum motivation, place the center of your monitor slightly above your eye line.

Creatures of Habit

Schachter's theory is not the only work linking the As If principle and eating.

I have worked at the University of Hertfordshire for many years. Throughout my time there, I have been fortunate enough to have brushed shoulders with many energetic and creative colleagues. Professor Ben Fletcher is one of those people.

Despite nearly always dressing in black, Ben is a happy man who shares my passion for quirky psychology that is relevant to everyday life. Ben's background is in business psychology, and much of his early work examines stress in the workplace. While he was carrying out this research, he discovered the downside of being a creature of habit.

Some people think and behave in highly inflexible ways. They might, for example, always try to solve problems using the same type of solutions, hold meetings in highly formulaic ways, and find comfort in the certainty that comes with their daily routines. In contrast, others embrace the unpredictable, enjoy lateral thinking, and are open to new ideas. Ben speculated that people who struggle to be flexible might be fine when they are working in an environment that is highly stable but struggle when faced with the need to change and adapt.

To find out if his hunch was correct, Ben created a questionnaire to measure a person's level of flexibility: "Do you sometimes behave in a way that your colleagues consider unconventional?" "Do you get bothered when people change plans at the last minute?" "You enjoy being presented with questions followed by a limited range of options: True or false?"* Ben then went to several different types of com-

---

* I made-up the last one.

panies and asked the employees there to complete the questionnaire, rate how well they coped with change, and indicate how anxious they felt. The results revealed that inflexible people did indeed find change difficult, tended not to enjoy their jobs, and were especially anxious.[34]

Ben then wondered whether the same concept also applied to people's lives outside their workplace. He speculated that many of the problems that people have in life are caused by their being inflexible and tied to certain habits. Overweight people are in the habit of eating too much and exercising too little. Smokers habitually reach into their pocket and pull out cigarettes. Many of those struggling to find a new relationship are in the habit of going to the same sorts of places and chatting to the same sorts of people. What would happen, wondered Ben, if these people were encouraged to behave as if they were not creatures of habit?

Muscle Magic

People who are highly motivated often tense their muscles as they get ready to spring into action. But is the opposite also true? Can you boost your willpower by tensing your muscles?

Ris Hung from the National University of Singapore and his colleague decided to find out.[35] Hung assembled several groups of participants and asked them to keep their hands submerged in an ice bucket for as long as possible, consume a healthy but terrible-tasting vinegar drink, or visit a local cafeteria and buy healthy food rather than sugary snacks. Each time, half of the participants were asked to tighten certain muscles by making their hand into a fist, sitting down and lifting their heels off the floor, holding a pen by tightly weaving it between their fingers, or contracting their bicep. Each of these exercises was designed to make the participants behave as if they were trying hard to exert self-control. The results showed that those carrying out the exercises were more likely to keep their hand in the bucket of ice for longer, down more vinegar, or buy healthier food.

Next time you feel your willpower draining away, try to tense a muscle. Consider, for example, making a fist, contracting your bicep, pressing your thumb and first finger together, or gripping a pen in your hand.

If all else fails, try crossing your arms. Another study, this one conducted by Ron Friedman and Andrew Elliot from the University of Rochester in New York, involved asking people to tackle difficult anagrams with either their arms crossed or resting on their thighs.[36] By folding their arms, people were acting as if they were persistent. The volunteers with their arms folded continued to persevere for nearly twice as long as those with their hands on their thighs.

To find out what happens when people are encouraged to alter their habits, Ben teamed up with another university colleague, Karen Pine, and together they developed a technique known as "do something different" (DSD). DSD consists of a series of exercises that encourages people to behave as if they have a flexible approach to life. They might, for example, be asked to stop watching television for a day, pen a poem, get in touch with an old friend, or travel to work using different routes. Over the years Ben and Karen have monitored the effects of these simple techniques on many different areas of people's lives.

Take, for example, Ben and Karen's work on weight loss.[37] In several studies they recruited people who wanted to lose weight and randomly allocated them to different groups. One of the groups was encouraged to carry out the DSD techniques for a month. These participants were not told to eat a healthier diet or get more exercise. Instead, they were encouraged to vary their ways of thinking and behaving, such as going to bed an hour earlier or turning off their cell phones for a day. In contrast, some of the other groups were either not given any instructions or followed a diet of their choosing.

After tracking the groups over several months, the results suggest that DSD helps people lose weight. Similar research has shown that the same technique can also help people stop smoking and increase their chances of finding a job.

Many undesirable behaviors, such as smoking and overeating, are the result of people behaving as if they are creatures of habit. By getting them to behave in a far more flexible fashion, they begin to see themselves in a completely different way. Suddenly they are no longer someone who just mindlessly repeats old behavioral patterns, but rather someone who is able to take control of his or her life and respond to the surroundings. It may seem like magic to this person. In reality, it is another example of the As If principle at work.

The As If principle provides a new and exciting way of looking at the thorny issue of motivation. It explains why rewards often fail and, more important, provides the basis for several motivational tools that are both quick to implement and highly effective. If you carry out the smallest of commitments, you are far more likely to engage with much bigger changes. Cross your arms, tense your muscles, and sit up straight, and you will persevere longer when the going gets tough. Behave as if you are no longer a creature of habit, and suddenly you will find it much easier to give up smoking and lose weight. These simple and quick techniques encourage you to change one small aspect of your behavior. In doing so, the As If principle springs into action and makes you feel like a new, and far more motivated, person.

Chapter 5

Persuasion

---

*Where we explore the problems of changing people's minds,
find out what really manipulates the masses, and discover how
cooperation can shape society*

"How can I tell what I think till I see what I say?"
　　　　　　　　　　—E. M. Forster, *Aspects of the Novel*

I. The Problems of Persuasion
II. Why Saying Is Believing
III. Manipulating the Masses
IV. On the Justification of Action
V. From Behavior to Bonding

I. The Problems of Persuasion

The Korean War saw two major conflicts. The first involved American and other pro-democracy forces fighting fierce battles with communist troops drawn from China and North Korea. The second was fought behind the barbed wire fences of North Korean's prisoner-of-war camps and was a battle for the hearts and minds of American service personnel captured during the fighting.

Hostilities officially came to an end in July 1953, with both sides agreeing to divide Korea into two separate countries. In January the following year, the prisoner-of-war camps were shut down, and the captured service personnel were released. It was only then that the full extent of the second conflict became apparent.

After the camps were closed, twenty-one American soldiers chose to remain in communist Korea, openly denouncing their own country and siding with an enemy that had killed more than thirty thousand of their comrades. In addition, a surprisingly large number of the American service personnel who did return home enthusiastically expounded the strengths of communism.

The family and friends of the servicemen who chose to remain in Korea were stunned, with one set of parents telling *Time* magazine, "I won't believe anything except that my boy wants to return home."[1] The world's media flocked to Korea to report the story and asked psychologists to explain the soldiers' seemingly incomprehensible decision. Some researchers suggested that the Koreans had brainwashed the American soldiers with flashing lights and white noise. Others speculated that they had used advanced forms of hypnosis or mind-altering drugs. They were all wrong.

Understanding exactly what actually happened to the American soldiers reveals how the As If principle can be used to change the world. Our journey begins by venturing deep into the psychology of persuasion.

"When I Read about the Evils of Drinking, I Gave Up Reading"*

Governments spend large sums of money trying to persuade the public to stop smoking, avoid excessive drinking, and eat a healthier diet. These well-meaning campaigns often rest on one assumption: tell people that their lifestyle is bad for them, and they will soon change their ways. Let the public know, for instance, that smoking causes cancer, and they will stop lighting up. Show them how alcoholism ruins lives, and they will curb their drinking. Point out how fatty foods clog up the arteries, and they will start eating more fresh fruit. There is, however, one small problem with this sensible approach: much of the time it doesn't work.

Irish stand-up comedian Andrew Maxwell recently made a television program in which he went on a road trip with five people who firmly believed in various 9/11 conspiracy theories. One of the believers, Rodney, was certain that the twin towers were not destroyed by two hijacked airplanes but instead collapsed because of a government-controlled explosion. Another member of the group, Charlotte, was equally sure that terrorists with such a limited amount of training wouldn't have been able to fly the planes into the twin towers.

Maxwell took Rodney and Charlotte to meet several experts who presented overwhelming evidence against their cherished theories. In one sequence, a demolition specialist showed just how difficult it would be to prepare charges to bring down buildings the size of the

---

* Attributed to comedian Henny Youngman.

twin towers. In another instance, a flight instructor demonstrated how easy it is to pilot aircraft. Did these experiences change Rodney and Charlotte's beliefs? Not at all. At the end of the program, Rodney and Charlotte remained unmoved by the evidence and explained that they continued to believe that the events of 9/11 were the work of the American government.

Similarly, in 1997, members of the Heaven's Gate cult convinced themselves that they would soon be transported away from earth on a spaceship traveling behind the Hale-Bopp comet. A few weeks before the comet was due to come close to the earth, some of the Heaven's Gate members went to a shop that sold scientific apparatus and purchased a sophisticated telescope. When the group looked through the high-powered lens, they could clearly see the comet but not the spacecraft. It would be reasonable to think that the experience might have caused the group to question their beliefs. Instead, they returned to the shop the following day, explained that the telescope was faulty, and asked for their money back.[2]

It would be nice to think that when it comes to the rocky relationship between evidence and belief, there is something rather special about Rodney, Charlotte, and the saucer-seeking members of the Heaven's Gate cult. Nice, but wrong. Although very few people think that the American government destroyed the twin towers or that spaceships lurk behind comets, we all hold other beliefs with similar levels of conviction. And when maintaining these beliefs in the face of disconfirming evidence, we are all capable of performing the type of mental gymnastics that conspiracy theorists and cult members display. Like them, we seek out the company of like-minded souls, avoid information that doesn't support our point of view, and question the integrity of those who dare to disagree with us. Despite our desire to be creatures of logic, we all find it surprisingly easy to ignore the facts if the facts don't fit our beliefs.

One study, for example, tracked the public reaction to a comprehensive scientific report pointing out the strong link between smoking and cancer. An impressive 90 percent of nonsmokers, compared to just 60 percent of smokers, said that they found the report convincing.[3] In another study, participants were first asked to say whether they were for or against an important issue, such as climate change.[4] Then everyone was presented with various arguments about the topic, with some of the arguments being especially plausible ("Climate change is most probably due to the greenhouse effect"), while others were extremely implausible ("Large numbers of scientists have been bribed to say that climate change is a reality"). The participants were asked to read each of the arguments and remember as many as possible. If they were rational human beings, they should have remembered a mixture of plausible and implausible arguments. In fact, a clear pattern emerged, with those on both sides of the debate remembering the plausible arguments that supported their own position and the implausible ones against it.

This "I have made up my mind, don't confuse me with the facts" approach creates a major hurdle for governments wanting to change hearts and minds. Print SMOKING KILLS in big black letters on packets of cigarettes, and smokers will still find a way of convincing themselves that lighting up can't be all that bad. Tell heavy drinkers about the horrors of alcoholism, and they will continue to believe that they will be fine. Run a campaign about the importance of healthy eating, and the obese will still scarf up impressive quantities of burgers and fries.

And this is just the tip of the iceberg.

On Saying and Doing

Psychologists have spent decades investigating the relationship between how people say they will behave and how they actually behave, not least the work of Leonard Bickman and his colleagues from Smith College in Massachusetts.[5] Bickman wanted to discover the connection between people's beliefs and behavior with something as simple as littering. He and his team found a busy main street and strategically placed crumpled-up pieces of paper several feet away from a trash can, ensuring that the paper was in the direct path of pedestrians. Next, the experimenters moved across the street and secretly recorded the percentage of pedestrians who picked up the litter and placed it in the bin. It turned out that the good folk of Massachusetts weren't too tidy, with only 2 percent of pedestrians picking up the paper and putting it in the bin.

In the next phase of the study, the experimenters stopped hundreds of the pedestrians after they had walked past the litter and asked them one question: "Should it be everyone's responsibility to pick up litter when they see it, or should it be left for the people whose job it is to pick it up?" What percentage of the pedestrians said that everyone should do their bit to keep the streets clean? Ten percent? Forty percent? Sixty percent? In fact, a remarkable 94 percent of the people who had just walked past the crumpled ball of paper said that they firmly believed that it was everyone's job to pick up litter.

Bickman's research suggested that with littering, people are highly skilled at a kind of Orwellian double-think that allows them to think one thing but behave in a completely contradictory way.

Eager to discover whether this strange inconsistency applied to other areas of life, researchers turned their attention to a variety of important topics, including morality. Would you describe yourself as a moral person? Someone who generally tries to do the right thing, to settle arguments in a fair way, and to behave in an ethical fashion? Faced with these types of questions almost everyone repeatedly checks the "Yes, that's me" box. But do they tend to actually behave in a moral and ethical way? Psychologist Daniel Batson from the University of Kansas decided to find out.[6]

Batson was interested in whether people who claim to be highly moral behave in a moral way, or whether they just like the idea of appearing moral but aren't so enthusiastic about the associated costs (a phenomenon that Batson has labeled "moral hypocrisy"). In one study, he first asked a group of people to rate how moral they were by answering several questions about their ethics. Did they believe in a just world? Did they generally try to do the right thing? Were they selfish or concerned for the welfare of others?

A few weeks later, Batson invited the same group of people to his laboratory one at a time and asked them to take part in an experiment. Each participant was told that the study involved a second person who was currently secreted in an adjoining room. He then explained that one of them would be given a raffle ticket that could result in their winning a large prize, while the other would spend thirty minutes adding up a series of numbers.

Next, Batson suggested that the allocation of the raffle ticket and number-adding task should be made on the basis of a coin toss and asked the participant if he or she thought this was a fair approach to the issue. When the participant agreed, Batson explained that if the coin landed heads up, then the person would get the raffle ticket and the person in the next room would be given the number list. If the coin landed tails up, the participant would be handed the number list and the person in the next room would get the raffle ticket.

Finally, Batson handed the coin to the participant and explained that she should step into the corridor, flip the coin, and then return to the laboratory and report whether it had come up heads or tails. He explained that he had no way of knowing the actual outcome of the coin toss and so was relying on the participant to tell the truth.

The results revealed something deeply strange. By chance, the coin should have come up heads half the time. However, 90 percent of the participants returned to the laboratory grinning like a Cheshire cat, explained that the coin had come up heads, and claimed the raffle ticket. In short, there was evidence that a large number of the participants were being a tad economical with the truth. Were the participants who had previously rated themselves as being highly moral more honest than the others? When push came to shove, even those who had previously claimed to hang out on the moral high ground couldn't bring themselves to tell the truth.

Batson's findings show that even for something as deep-seated and important as our sense of morality, beliefs often do not predict behavior.

Manufacturing a Lack of Consent

The findings from the research on littering and morality are not an exception. Time and again, psychologists have discovered that people are highly skilled at holding one belief but behaving in a completely contradictory way.[7] Given the problems associated with changing beliefs and behavior, it isn't surprising that many government campaigns have struggled to make a difference, and a good example is the Hutchinson Smoking Prevention Project.[8]

In the late 1980s and early 1990s, the U.S. National Cancer Institute spent around $15 million creating and assessing a large-scale campaign designed to prevent children from smoking. In this part experiment and part public education program, more than four thousand children from twenty randomly chosen school districts in Seattle were bombarded with information designed to prevent them lighting up. For months these children attended special lessons and were given all sorts of helpful antismoking advice, including how to resist peer pressure and ignore tobacco advertising. Another four thousand children in twenty different districts weren't given this information and so acted as a control group.

The researchers then tracked down the vast majority of the children two years after they had left high school and asked them whether they smoked. Worried that the children might not tell the whole truth about their smoking habits, the investigators even went to the effort of measuring the amount of nicotine-based chemicals in the teenagers' saliva. The results were as dramatic as they were disappointing. Had the antismoking campaign produced the intended effect? Of the children who came from the districts that had run the campaign, 29 percent smoked versus 28 percent of children from the control districts. Spending millions in an attempt to prevent children from smoking had had almost no impact.

Unfortunately this is not an isolated example. Another American national antismoking campaign encouraged parents to try to dissuade their children from smoking using the slogan "Talk. They'll Listen." The result? Children displayed an age-old tendency to disobey authority figures, with the campaign ads making them less likely to believe that smoking was dangerous and more likely to say that they intended to smoke ("Don't talk. Otherwise they'll do the complete opposite").[9] In Britain, the Department of Health spent more than 3 million pounds encouraging the public to eat five portions of vegetables a day, only to see vegetable consumption drop by around 11 percent.[10] And between the late 1990s and 2004, the U.S. Congress gave almost $1 billion to fund an extensive antidrug media campaign, only to discover that the ads had failed to persuade teenagers to avoid smoking marijuana and might even have actively encouraged some to try it.[11]

Realizing that the conventional approaches to changing hearts and minds frequently failed to produce results, researchers started to examine other ways of transforming attitudes and beliefs. Eventually a young psychologist fresh out of graduate school came up with a radical idea that changed the entire course of behavioral science.

Manufacturing Consent: Part One

Please complete the following questionnaire by giving each statement a rating between 1 (strongly disagree) and 5 (strongly agree). Jot down your answers on a piece of paper.

Statement	Your rating from 1 (strongly disagree) to 5 (strongly agree)
1. I leave the tap running while I brush my teeth.	
2. I catch a flight when I could have traveled by car or train.	

3. The light bulbs in my house and office are not energy efficient.	
4. I don't place my trash in recycling bins.	
5. I buy new goods rather than second-hand ones.	
6. I leave lights on when I leave a room.	
7. I strongly support the idea of green living.	

Many thanks. More about this later.

II. Why Saying Is Believing

Around the turn of the last century, sociologist William Graham Sumner argued that certain beliefs are biologically ingrained in the human brain. Sumner referred to these beliefs as "folkways" and declared that they were extremely difficult to change.

In 1896 the U.S. Supreme Court was required to rule on the legality of racial segregation. Many of those in favor of segregation argued that the notion of one race being superior to another constituted one of Sumner's folkways, and thus any attempt to legislate against it would be futile. The Supreme Court was swayed by the argument and, citing the maxim that "law-ways cannot change folkways," ruled that all American citizens should have access to the same public services but that there could be separate facilities for each race. In reality, the facilities available to African Americans were usually of a much lower quality than those available to others.

From the middle of the 1940s onward, the American civil rights movement worked hard to overturn the racial segregation laws, and in the early 1950s the Supreme Court was required to reexamine the legality of segregated schools. Lawyers in favor of desegregation argued that the 1896 separate-but-equal doctrine was unconstitutional, in part because it created a sense of inferiority among African American children. The legal teams supported their case by citing the work of several behavioral scientists, including psychologists Kenneth and Mamie Clark.

In a series of now-classic studies, the Clarks had asked African

American children to select a white or a black doll and then describe the chosen doll's character.[12] Almost all of the children preferred the white doll and attributed positive characteristics to it. The Clarks thus argued that the results provided a vivid demonstration of how segregated schools caused African American children to suffer from low self-esteem. The arguments proved persuasive, and in 1954 the Supreme Court unanimously ruled that separate educational facilities were unconstitutional. Other similar rulings quickly followed, as the courts came down against racial segregation on buses and other forms of public transportation.

At the time of the rulings, social psychologist Daryl Bem was a graduate student at the University of Michigan. Although he had originally intended to study physics, Bem had become fascinated by the impact of the civil rights movement on public belief and switched to psychology. He decided to analyze the results of surveys examining the attitudes of white Americans toward racial segregation before and after the 1954 Supreme Court ruling. His analysis soon revealed a rather curious pattern.

Before the landmark decision, only a minority of white Americans supported desegregation. For instance, in a survey undertaken in 1942, only 30 percent of white Americans were in favor of school integration, 35 percent approved of residential integration, and 44 percent supported integrated transportation. However, just two years after the Supreme Court's decision, the numbers had significantly increased, with a 1956 survey showing 49 percent now supporting school integration, 51 percent approving of residential integration, and 60 percent favoring integrated transportation.[13]

For years the American civil rights movement had struggled to gain public support for desegregation. Now, within just a few years of the Supreme Court ruling, far more white Americans supported the idea than ever before. Eager to explain this effect, Bem searched through various psychology textbooks and eventually came across William James's work on behavior and emotion.

As we discovered in Chapter 1, according to the As If principle, behavior causes emotion. When, for example, people smile, they feel happy, and when they frown, they feel sad. Bem wondered whether the principle does not just determine how people feel, but also affects what they believe. Common sense suggests that thoughts create behavior. Imagine, for example, that you want to go out one evening and have the choice of seeing a movie or a play. You know that you prefer movies to plays and so head for the movie theater. In this example, your thought ("I prefer movies to plays") caused your behavior (going to the movie theater). Following in James's footsteps, Bem turned this commonsense view of the human psyche on its head and suggested that behavior influences what people believe. So if, for instance, you want to go out for the evening and are subtly persuaded to go to the theater to see a play, you would look at your behavior and unconsciously think, "Hold on! Here I am watching a play. I guess I must enjoy the theater more than the movies." As a result, you would end up feeling surprisingly positive toward live theater (see the diagram).

Common sense suggests that the chain of causation is:
I like movies— Go to the movie theater
The As If theory suggests that the reality is:
Go to the movie theater— I must like movies

This extension of the As If principle into the heady world of thought control could explain why the Supreme Court's ruling about desegregation caused a sea change in public belief. The legal ruling required people to behave as if they supported desegregation, which in turn caused them to unconsciously think, "Hold on! Here I am behaving as if I support desegregation. I guess I must believe that racial equality is a good idea." As a result, they then developed a new and much more positive view about desegregation.

Although the surveys before and after racial desegregation supported the As If principle, they didn't provide conclusive proof: the seismic shift in public opinion could have been due to other factors, such as the extensive campaigning by the civil rights movement. Eager to discover whether the As If principle really did affect what people believe, researchers retreated to their laboratories and conducted a series of systematic experiments.

Comment and Commitment

The Vietnam War cost the lives of more than fifty thousand American soldiers. To bolster public support during the war, the American government frequently attempted to put a positive spin on the conflict, arguing that the Communist North of the country would soon be defeated or that the South would soon become sufficiently strong to defend itself. President Lyndon Johnson was responsible for sending more troops to the country than any other leader. From time to time he became aware that members of his administration were privately expressing concerns about the war. Did Johnson pull these politicians into his office and try to explain his point of view? Not at all. Instead, he used a far more unusual tactic: he sent the doubters on a fact-finding mission to the country along with a group of reporters.[14]

Johnson was well aware that members of his administration would be unlikely to express their private doubts in public and would instead be forced to give high-profile speeches defending the government's policies. The As If principle predicted that the doubters would hear themselves giving speeches supporting the president and would eventually come to believe their own remarks.

Researchers used the same approach in their laboratory tests examining whether the As If principle influenced what people believed. To find out, researchers invited participants into the laboratory and asked them to complete questionnaires about their political beliefs.[15] Next, half of the participants were asked to give a short speech in which they argued in favor of a political party that they opposed, while the others watched these speeches through a two-way mirror.

159

Two weeks later, all of the participants completed another question-naire about their views.

The As If principle predicts that those who had given the speeches would have seen themselves supporting a political party, and so come to believe that perhaps this party wasn't so bad after all. In contrast, those just hearing the speeches would have received exactly the same information, but because they didn't see themselves arguing the point, would not have changed their beliefs. The results supported the theory, with just a few minutes of role play achieving what a constant bombardment of campaigning and political ads had failed to do.

Over the years, the same procedure has been used in many con-texts. People have been filmed presenting talks in favor of a whole host of issues such as abortion, the dangers of drunk driving, and greater police powers. On each occasion, behaving as if they be-lieved a certain argument achieves what a hundred rational reasons couldn't, quickly changing their attitudes in favor of the position they were asked to support.[16] Indeed, the depth of change is such that the participants often deny ever holding their original opin-ions. If they are shown their original questionnaires, they argue that the forms have been faked or claim that they misread the ques-tions.[17]

This mechanism also accounts for many otherwise inexplicable shifts in belief. At the start of this chapter, I described how, at the end of the Korean War, a large number of U.S. servicemen decided to remain in Korea, and many of those who did return home were en-thused about the benefits of communism. Extensive interviews with those who had survived the harsh conditions of the camps revealed that these beliefs had not been instilled with the help of hypnotism, drugs, or physical punishment.[18] Instead, the Chinese authorities had deliberately employed the As If principle to alter the minds of the American prisoners of war.

This unusual form of mind control often started the moment the prisoners entered the camps, with the guards shaking the hands of the new detainees and saying, "Congratulations, you've been liberated." During the next few weeks, the prisoners were required to attend long lectures on the benefits of communism and then asked to discuss the talks in small groups. A member of the Chinese Communist party was often assigned to each group to help ensure that the prisoners reached the "correct" conclusions. If anyone in the group openly displayed any anticommunist sentiments, then all of the prisoners had to endure the lecture and discussion again.

Shortly after the servicemen were captured, the Chinese guards asked them to jot down a few short pro-communist statements—for example, "Communism is wonderful," "Communism is the way of the future," and "Communism is the most enlightened form of government." Many of the Americans were happy to oblige because the request seemed so trivial and compliance was often rewarded with a small bar of soap or some cigarettes. A few weeks later, the guards upped the ante and asked the prisoners to read the statements aloud to themselves. Once again, most agreed to do so. And then a couple of weeks after that, the Americans were asked to read the statements out loud to their fellow prisoners, and finally to engage in mock debates arguing why they believed the statements were correct.

In addition, highly valued goods such as fresh fruit or candy were offered to any soldiers who were prepared to write pro-communist essays for the camp newsletter. When the essays were published, the authors were asked to wear a Mao Tse-tung badge and were excused from unpleasant camp chores. Many of the prisoners were again happy to oblige.

Over time such behavior caused many of the American prisoners to change their attitudes toward communism, and even persuaded some to stay in Korea rather than return home. The As If principle helps explain this dramatic change in belief. The Chinese did not have to resort to physical punishment or arcane brainwashing techniques. All they had to do was ensure that the prisoners saw themselves repeatedly saying that they supported communism and then leave them alone to develop beliefs that were consistent with their comments.

The same effect can be used to influence entire populations.[19] Saying "Heil Hitler" every day would have encouraged many ordinary Germans to become more open to Nazi ideology. Having people repeatedly sing a national anthem will make them more patriotic. Making children pray every morning will increase the likelihood of their adopting religious beliefs.

In each instance, saying becomes believing. Intrigued by this research, some people started to examine whether other types of behavior also had the power to persuade. Perhaps the two best-known examples focused on eye color and the creation of the Third Wave.

One Finger, One Thumb, Keep Moving

The As If principle has the power to radically change people's ideological beliefs. The same behavior-creates-belief process can also be used to shape people's thoughts about many aspects of their everyday lives.

It's time for a quick experiment. Hold out your thumb as if you are giving a thumbs-up sign and then read the following paragraph:

Donald was faced with a difficult situation. For the past few months, he had been renting an apartment but now wanted to move out. His lease had expired, but his landlord was refusing to return his deposit. After repeatedly asking for his money back, Donald became angrier and angrier. One day, he finally lost his temper, picked up the telephone, and shouted a stream of abuse at his landlord.

What do you think of Donald? Do you approve of his behavior in this particular instance? Now extend your middle finger as if you are giving someone the "bird," and then read the paragraph again. What do you think of Donald and his behavior this time?

In most Western countries, raising your middle finger toward someone is usually a sign that you don't like that person, whereas giving a "thumbs up" is a far more positive signal. In each instance, whether you like or dislike a person influences your behavior. But could the opposite also be true? Could your gestures change the way you think about a person?

This mini-experiment is based on a study conducted by Jesse Chandler from the University of Michigan.[20] Chandler invited a group of participants into his laboratory and explained that they were going to take part in an experiment about gestures and language. The participants were first asked to either extend their middle finger or hold out their thumbs and then read about Donald and his landlord. At the end of the story, participants were asked to rate how much they liked Donald. When they read the story with their middle finger raised, they thought that Donald was an aggressive man. In contrast, when they read the story with their thumb raised, they thought Donald was far less aggressive and much more likable.

Two key implications flow from the work. First, on a theoretical level, it demonstrates how just a few seconds behaving as if can influence what you think about someone. Second, on a more practical level, if you're struggling to get along with a work colleague, try giving this person a thumbs-up on a regular basis.

When it comes to everyday persuasion, this is just the

tip of the iceberg. In another study, for example, students were asked to listen to a discussion suggesting that their college tuition should be increased.[21] Some of the students were asked to nod their heads up and down while they listened to the discussion (causing them to nod as if they agreed with the arguments) while others shook their heads from side to side (shaking their heads as if they disagreed). The students were then asked how much their tuition should be. Those who had been shaking their heads from side to side produced much lower estimates than those who had been happily nodding away. Want to encourage someone to agree with you? Then subtly nod your head as you chat. The other person will reciprocate the movement and be strangely attracted to your way of thinking.

Then there is the issue of chairs. In another study, experimenters had participants sit on either a hard wooden chair or a soft cushioned chair and then asked them to role-play negotiating for a new car with a stranger and rate the personality of the stranger. Those in the hard chairs were more inflexible in their negotiations and saw the strangers as less likable. In short, the evidence is that hard furniture creates hard behavior, emphasizing the importance of soft furnishings in your home and office.

III. Manipulating the Masses

In the late 1960s Jane Elliott was working as a primary school teacher in Riceville, Iowa. On April 4, 1968, Martin Luther King Jr. was assassinated, and Elliott decided to hold a class discussion about racism. She was disappointed with the impact of the lesson and began to think about other ways of getting her pupils to engage with the topic. That evening she devised a daring plan.[22]

The following day, Elliott told her class that children with blue eyes were superior to children with brown eyes. At first many pupils were skeptical, and so the quick-thinking Elliott created some pseudoscience to support her claim, explaining that blue eyes were the result of melanin and that research had shown that the same chemical was linked to higher intelligence.

The majority of the pupils bought the pseudoscience, and she proceeded with the next phase of the exercise. Because the blue-eyed children were superior, explained Elliott, they would be given special privileges, including second helpings at lunch, extra-long breaks, and the opportunity to sit at the front of the classroom. In contrast, the brown-eyed children were treated like second-class citizens. They were encouraged to play only with other brown-eyed children and not allowed to drink from the blue-eyed children's water fountain. To make the difference between the blue-eyed and brown-eyed pupils especially obvious, Elliot also had each group of children wear different-colored scarves.

Suddenly the As If principle kicked in, and the enforced changes in behavior resulted in a dramatic shift in the children's personalities. The blue-eyed children became arrogant and bossy, while the brown-eyed children became timid and subservient. As a result, the blue-eyed children started to outperform the brown-eyed children on various tests.

A few days later, Elliott told her class that she had made an error, and that actually it was the brown-eyed children who were superior to the blue-eyed children. Suddenly the children's sense of identity flipped, with the blue-eyed children becoming much more reserved and the brown-eyed children acting in a far more assertive way. On the final day of the exercise, she revealed that there was no difference between blue-eyed and brown-eyed children, explained that the exercise had been designed to help the class understand what it feels like to be the subject of discrimination, and told the children to take off their scarves. Many of Elliott's children cried and hugged one another.

The media caught wind of Elliott's exercise, and she was eventually invited on Johnny Carson's *Tonight Show*. Although viewers across America were touched by the story, many Riceville residents thought that the item gave the impression that their town was a hotbed of racism. As a result, many of Elliott's colleagues refused to speak to her, and her family was subjected to verbal and physical abuse.

Unperturbed, Elliott repeated the exercise many times over the following decades. Time and again, the same results emerged, with the children's behavior quickly affecting their beliefs about one another. As adults, many of those who took part in Elliott's exercise describe how the experience completely changed the way they viewed the disenfranchised and powerless in society.

Elliott eventually left teaching in the mid-1980s to become a full-time diversity trainer."[23]

Around the same time that Elliott was handing out privileges on the basis of eye color, another teacher was attempting to use the same technique to create a microcosm of Nazi Germany. In 1967, a charismatic twenty-five-year-old history teacher and basketball coach named Ron Jones was working at a high school in Palo Alto, California.[24] Ever ready to explore new methods of teaching, Jones decided to employ an unusual hands-on approach to explaining some of the factors that contributed to the formation of Nazi Germany.

At the start of one of his lessons, Jones lectured about the beauty of discipline and self-control. In order to emphasize his point, Jones had his students repeatedly sit up straight, put their feet flat on the floor, and place their hands flat across the small of their backs.

On the following day, Jones spoke about the value of community and had his class repeatedly recite the phrase "Strength through Community." At the end of the lesson Jones also created a class salute, explaining how his students needed to bring their right hand up to their right shoulder in a curled position. When the end-of-lesson bell rang, Jones slowly carried out the action and saw his entire class return the salute.

The next day Jones gave each of his students a membership card and asked them to help enlist other students into a newly formed organization he called the "Third Wave." Every member was also encouraged to report anyone openly expressing skepticism about the project.

Word about the Third Wave quickly spread throughout the school, with some students creating banners and leaflets to promote the cause. Soon Jones's organization boasted more than a hundred members, with many students developing highly authoritarian characteristics and demanding strict obedience of the rules. Jones decided that his experiment was getting out of hand and decided to call an end to the project. He announced that all members of the Third Wave should meet in the school's auditorium for a special announcement.

More than two hundred students gathered at the appointed time, with many wearing white shirts and home-made armbands. Jones turned on a projector and showed the group images depicting the history of the Third Reich and the Nuremberg rallies. As the final frame faded, Jones announced that the exercise was designed to show how easily people's behavior and beliefs can be manipulated and to emphasize the need for everyone to take responsibility for their actions. Many of the students started crying as the reality of the situation hit them.

A few years after Jones conducted his experiment, he was refused tenure and has since spent more than thirty years writing, speaking, and working with people with mental disabilities. Jones's work was subsequently described in the novel *The Wave*, which became required reading in many German schools. In 2008 the experiment acted as the inspiration for the German movie *Die Welle*, and in 2010 Jones staged a musical about his project in San Francisco.

Manufacturing Consent: Part Two

Complete the following questionnaire by giving each statement a rating between 1 (strongly disagree) and 5 (strongly agree). Jot down your answers on a piece of paper.

Statement	Your rating from 1 (strongly disagree) to 5 (strongly agree)
1. I don't leave the tap running while I brush my teeth.	
2. I rarely catch a flight when I could have traveled by car or train.	
3. Few of the light bulbs in my house and office are not energy efficient.	
4. I place my trash in recycling bins.	
5. When I can, I buy second-hand goods rather than new ones.	
6. I don't leave lights on when I leave a room.	
7. I strongly support the idea of green living.	

Look at the answer that you have just given to Question 7. Next, look at the answer that you gave to the same question in part one of the exercise. According to research conducted by psychologist Shelly Chaiken from New York University, it is likely that you gave a lower rating the first time in this exercise.[25]

Chaiken's work shows that the relationship between the As If principle and belief also applies to the way in which people think about the past. The questionnaire in the first part of the exercise asked about times when you *didn't* act in an environmentally friendly way by, for example, not turning off the tap when you brush your teeth or took a plane when you could have gone by train. In contrast, the questionnaire in the second exercise asked about times when your behavior was much more green, such as when you recycle or turn off the lights when you leave a room.

The As If principle causes people to complete the first questionnaire and think, "Well, I don't behave as if I am an environmentally friendly person, so I don't support green living," while the second questionnaire makes them think, "Well, my behavior seems very pro-environment, so I must hold green views."

By reminding people of specific aspects of their past and present behavior, questionnaires like this can be used to manufacture, rather than simply measure, belief.

IV. On The Justification of Action

In 2004 the television program *60 Minutes* broadcast a shocking report about the terrible abuse being inflicted on prisoners held at Iraq's Abu Ghraib prison. According to the report, American soldiers had carried out a range of physical and psychological abuse, including beating, rape, and torture. The world was shocked to see photographs of prisoners being pulled around like dogs, being led to believe that they were about to be electrocuted, and being piled naked on top of each other in prison corridors. The U.S. Department of Defense responded by removing several soldiers from duty, with many of them subsequently being charged with a range of offenses. One question weighed heavily on the public's mind: How did the soldiers come to commit such terrible atrocities?

A key part of the answer centers around the As If principle.

One of Aesop's best-known fables concerns a fox and a bunch of grapes. According to the story, a fox is strolling through an orchard when he happens to see a lovely-looking bunch of grapes hanging from a lofty branch. Realizing that he is thirsty, the fox draws back a few paces and takes a running jump at the grapes. Unfortunately he misses the bunch. Never one to give up easily, the fox tries a second jump, but once again misses the grapes. Throughout the afternoon, the fox repeatedly attempts to reach the branch and fails each time. Eventually the fox gives up, walks away from the orchard empty-mouthed, and convinces himself that he doesn't want the grapes because they are probably sour.

In addition to providing the basis for the popular idiom "sour grapes," the story provides a perfect illustration of the As If principle. The fox started off believing that the grapes looked great, but when he was forced to leave the orchard without them, he formed a new and negative view of the grapes. In short, the fox first looked at his actions and then created a new belief to justify them.

A small group of researchers decided to discover whether the same process influences people's beliefs. Would they, for instance, decide that they disliked objects that they couldn't obtain and become especially fond of those that were within their grasp?

In one series of studies, participants were first asked to rate how much they liked several products, such as a coffee maker, sandwich grill, toaster, portable radio, and so on.[26] Next, the experimenters selected two items that had received the same rating, showed them to the participant, and asked the person to choose one of the items as a gift. The chosen item was then put into a box, securely tied with string, and placed next to the participant's coat. All of this was designed to give the participant the impression that she would be taking the object home with her (in reality the experimenters were on a tight budget and so reclaimed their "gift" the moment the study had finished). Finally, the participant was asked to rate the desirability of both items a second time.

Before making her decision, the participant had indicated that both objects were equally appealing. However, according to the As If principle, the moment people behave as if they like one item more than the other, they justify their actions by convincing themselves that they are especially fond of the chosen object. The results confirmed the theory: the participants suddenly found their chosen item significantly more desirable than before.

In another study, researchers left their laboratories and headed for the racetrack.[27] They randomly selected a group of people who were just about to place their bets and asked them to rate the chances that their horse would win the race. On average people said that their horse had a "fair" chance. The research team then approached another group of people seconds after they had placed their bet and asked them the same question. According to conventional theories of belief, people's confidence should be roughly the same before and after the bet. However, the As If principle predicts that people unconsciously think, "Well, I have just seen myself placing my bet, so I must be pretty certain that my horse will win," and so feel more confident about their decision. As the researchers discovered, participants who had just handed over their betting slips now rated their chosen horse as having an "excellent" chance of being first past the post.

The Power of Warm

From an early age, we associate the feeling of warmth with safety and security (think hugs and open fires), and coldness with unfriendliness (think "getting the cold shoulder" and "icy stare"). Inspired by this notion, psychologist Chen-Bo Zhong from Northwestern University decided to discover whether being excluded from the group really does make people feel a little chilly.[28] In one experiment, Zhong gathered together a group of people and asked half of them to think about a time in their lives when they had felt rejected by others and the other half to remember an experience in which they had been accepted as part of a group. Everyone was then asked to estimate the temperature of the room that they were in. Remarkably, those who had simply thought about a time when they were alone rated the room as significantly colder than those who had imagined themselves being part of a group. It seems that loneliness does indeed make people feel physically cold. Zhong believes that this association between warmth and social inclusion may start early in life, with children who are hugged by their parents feeling a sense of belonging and physical warmth.

Given that lonely people feel physically cold, the As If principle predicts that warming people up should make them feel far friendlier. Research conducted by University of Colorado psychologist Lawrence Williams suggests that this is indeed the case.[29] Williams conducted an experiment in which participants were handed either a hot cup of coffee or a cold drink, asked to read a short description of a stranger, and then asked to rate the stranger's personality. The participants who had been warmed up by the coffee thought that the stranger seemed much friendlier than those who had been clutching the iced drink.

The implications are obvious: if you are trying to befriend someone, abandon the idea of frozen cocktails in an air-conditioned bar and instead opt for a steaming mug of tea or coffee in front of a roaring fire.

This effect emerges in many other situations. Imagine, for example, going shopping for a winter jacket and seeing an entire rack of great-looking coats. All of the jackets are very attractive, and it takes you an inordinate amount of time to decide which one to buy. However, the instant you pass your credit card to the cashier, you suddenly start to justify your behavior by thinking about all the reasons your coat is obviously so much better than the ones you left on the rack. Within moments, your behavior has caused you to form a new belief, and now you become certain that you have made the right decision. Unfortunately, the effect can create unhealthy levels of overconfidence, causing politicians to doggedly pursue failing policies, companies to continue promoting unsuccessful products, and investors to continue to support poor initiatives.

The As If principle does not just explain how people's behavior can cause them to become overconfident. It also accounts for some of the surprising effects that can emerge whenever people decide to do something that deep down they really don't want to do.

In the 1960s psychologist Jack Brehm from Duke University carried out work exploring whether the As If principle can be used to change the way children view vegetables.[30] Brehm first asked about fifty children to rate the degree to which they liked each item on a long list of vegetables. A few weeks later, Brehm told the children that he was interested in whether their perception of vegetables would change after they had eaten them and asked them whether they would mind tasting a randomly selected vegetable. In reality, the vegetables were not randomly selected. Instead, Brehm had gone shopping and bought the very vegetables that each of the children said they disliked. Each child was presented with a serving of the despised vegetable and asked to try to eat three portions of it each week for the next few weeks. A month later, Brehm tracked down the children, presented them with the long list of vegetables, and again asked them to rate each of the items. The As If principle predicts that the children would have seen themselves eating the vegetables and justified their behavior by convincing themselves that they must quite like the food after all. And this is what happened. The message from Brehm's study is clear: persuade someone to do something he doesn't enjoy and he will often end up justifying his actions by convincing himself it is not so bad after all.

This curious phenomenon helps explain why legislation often results in a significant shift in public opinion. When smoking in public places was banned in Britain, many smokers found it much harder to light up and so came to adopt an antismoking attitude. Similarly, when the British government passed a law that made wearing a seat belt compulsory, opinion polls showed that many people started to believe that buckling up was a good idea. In each instance, as the As If principle suggests, people's behavior caused them to adopt certain beliefs.

However, whereas some of these changes can be positive, the same idea can also cause pain and suffering.

A few years ago, psychologist David Glass from Ohio State University conducted a remarkable experiment.[31] He invited people to his laboratory one at a time and introduced them to another participant (actually a stooge working for the experimenters). The participant and stooge were allowed to chat for a few minutes and then the participant was asked to answer a few questions about her newly found friend, including whether she would admit the person into her circle of close friends and whether she would like to share an apartment with this person.

Next, honoring a long-held tradition in social psychology experiments, the experimenter explained that the two participants would now carry out a study in which one of them would attempt to learn a long list of words and the other would give an electric shock every time the other person made a mistake. The experimenter flipped a coin to assign the two tasks and, yes, you guessed it: the genuine participant was asked to administer the shocks. The participant then went into the room containing the controls to the shock machine, and the stooge went into an adjoining room.

The two rooms were connected with an intercom that allowed the participant to hear the stooge. Each time the stooge made a mistake (which turned out to be surprisingly frequent), the participant thought that she was delivering a 100-volt shock (in reality, the machine was disconnected, and the stooge was quietly sitting next door enjoying a sandwich).

After administering several shocks, the participant was asked how she felt about what she had done and again rate the degree to which she liked the stooge. The participant could have concluded that she was a bad person or just following orders. However, most participants were reluctant to think badly of themselves and justified their behavior by saying that their fellow participant was not such a nice person after all and so was deserving of the shocks. As the As If principle predicts, the participants formed a new belief on the basis of their behavior. In this situation, they behaved as if they didn't like a person and so ended up believing that that person was not likable and deserving of punishment.

At the start of this section, I described how American soldiers abused prisoners in Iraq's Abu Ghraib prison. Glass's research illustrates how the As If principle helps explain the atrocities. If a guard with high self-esteem administers a small but illegal punishment to a prisoner, the guard may convince himself that the prisoner is a bad person and so deserving of such treatment. This may then cause the guard to believe that he is justified in carrying out more abuse, which in turn causes him to believe that the prisoner is deserving of even greater punishment. If this process remains unchecked, over time it can feed on itself, resulting in a situation that quickly spirals out of control and creating the type of shocking behavior seen in Abu Ghraib.

Fortunately, not all of the research into how the As If principle affects what people think is so gloomy. On a more positive note, other research has shown how the principle can also be used to bond people together and even save lives.

Charisma, Empathy, and the As If Principle

Rate the degree to which you agree with the following statements by assigning each of them a rating between 1 (strongly disagree) and 5 (strongly agree). Jot down your answers on a piece of paper.

Statement	Your rating
1. I often find myself tapping my feet along to music.	
2. I feel sad when I see someone who looks lonely.	
3. I enjoy hugging people.	
4. I care a great deal about animals.	

5. I find it easy to make others laugh.	
6. I quickly become anxious if people around me become nervous.	
7. I find it very easy to catch someone's eye.	
8. I often cry when I see a romantic movie or hear a love song.	
9. People often describe me as the life and soul of the party.	
10. I enjoy giving people presents and looking at their faces when they open them.	

Ask people to name a charismatic figure, and they are likely to come up with the likes of Martin Luther King Jr., Nelson Mandela, John F. Kennedy, and Barack Obama. However, ask them to explain why these people possess the X-factor, and they will struggle to define this elusive quality.

We all mimic the facial expressions and body language of those around us.[32] This process happens unconsciously, automatically, and in the blink of an eye. When you see someone smile, the sides of your mouth start to move toward your ears. Similarly, if you are faced with a frown, your brow begins to furrow. This process allows emotions to spread from one person to another and has evolved to promote group empathy and cohesion.

Some people are naturally skilled at using their face, body, and voice to induce their own emotions in others. Research conducted by psychologist Howard Friedman from the University of California shows that such people are seen as highly charismatic.[33] They are able to induce in those around them the passion and energy that they feel, often initiating a wave of emotional contagion that spreads from one person to the next. This process can make an entire room feel highly energized and transfix an audience of thousands. Charismatic communicators bypass the normal paths to persuasion, making people feel rather than think, and so speak directly to the heart.

Similarly, other people will be especially good at "catching" other people's emotions. In one experiment, Per Andréasson from the Uppsala University in Sweden first asked a group of participants to rate how empathic they were and then showed them photographs of people who were either happy or angry.[34] When the highly empathic participants saw a happy face, the muscles around the sides of their mouth suddenly burst into action. In contrast, the participants who had indicated that they had a low sense of empathy showed almost no reaction. Similarly, when the highly empathic participants saw an angry face, their eyes instantly started to narrow, whereas the low empaths kept a straight face. By behaving as if they are experiencing the emotions of those around them, highly empathic people literally feel the pain and joy of others.

The questionnaire at the start of this section measures the degree to which you are able to send and receive emotions.[35] To discover your charisma score, add up your responses to the odd-numbered questions. To find out your empathic score, add up your responses to the even-numbered questions.

Charisma score: _____
Scores between 5 and 15 are classified as low, scores between 16 and 25 as high.

Empathy score: _____
Scores between 5 and 15 are classified as low, scores between 16 and 25 as high.

V. From Behavior to Bonding

Muzafer Sherif was born in Turkey in 1906. When he was a teenager, he witnessed some of the atrocities that the Greek army carried out during the Greco-Turkish War. Appalled at seeing innocent Turkish citizens robbed, tortured, and killed, Sherif wanted to understand why people sometimes behaved in such savage ways. He enrolled for a degree in psychology, then emigrated to America to continue his education at Harvard University and eventually went on to enjoy a highly successful academic career.

During his time in America, Sherif conducted a highly controversial experiment to help explain some of the terrible scenes that he had witnessed as a teenager.[36] In doing so, he also inadvertently carried out an impressive test of whether the As If principle can be used to move people emotionally closer to one another.

Sherif first needed to find some unsuspecting volunteers to take part in his experiment. He hung around several school playgrounds, secretly observing groups of twelve-year-old boys ("And that, m'lord, rests the case for the defence"). He was on the lookout for boys who seemed psychologically stable, reasonably popular, and of normal intelligence. Whenever Sherif saw a boy who seemed to fit the bill, he scurried away and checked the boy's school reports to ensure that he was not prone to temper tantrums and had a good attendance record. Over time, he assembled a list of possible pupils and moved on to the second stage of the selection process. Sherif arranged to meet the boys' parents, explained what he was up to, and asked whether he could whisk their son away for a three-week psychology experiment.

179

Eventually he assembled a group of twenty-two boys. None of the boys knew they were about to take part in a research project. Instead, they were told that they had been selected to attend a summer camp.

In the next step, Sherif built an artificial world in which he could manipulate and monitor his carefully chosen group of boys. After checking several possible locations, he eventually came across an isolated state park in Oklahoma. Located more than forty miles from the nearest town, the park consisted of two hundred acres of forest. Hidden from civilization, it was the perfect place for Sherif's project.

The park housed two separate camps hidden from each other by dense greenery. Each camp consisted of cabins, a dining hall, a swimming pool, and a lake for boating. The camps also had shared access to a large baseball field.

Sherif randomly allocated each of the boys to one of two groups, ensuring that neither group knew about the existence of the other, and then bused one of the groups to the first camp and the other group to the second camp.

Throughout the experiment, Sherif played the role of the park janitor, and his research team acted as camp coordinators. Although they all often appeared disinterested in what was happening, in reality Sherif and his team made copious notes about the boys' daily behavior, secretly recorded their conversations, and took more than a thousand photographs.

During the first part of the experiment, Sherif wanted each group of boys to bond together and so arranged for them to participate in various joint activities, including hiking, playing baseball, and swimming. He also asked each group to come up with a name for their gang and create their own flag. One group labeled themselves the "Rattlers," and the other went with "Eagles."

Sherif's plan worked. Within just a few days, each group of twenty-two strangers had become a closely knit community. Pleased with the progress, he moved on to the second phase of the study: the experimental production of hatred.

On an agreed morning, the researchers told the Rattlers about the existence of the Eagles and the Eagles about the Rattlers. Both groups had enjoyed playing baseball, with each group thinking that it had sole access to the baseball field and so considered it part of their territory. The researchers decided to use this situation to create a sense of competition and told the Eagles that the Rattlers had also been using the field and vice versa. Both groups felt threatened and said that they wanted to challenge the other group in some type of competition. The researchers suggested that they organize a tug of war or a game of baseball and offered to reward the winning team with medals and trophies.

The following day, both groups agreed to a game of baseball. It was a testosterone-charged affair right from the very start. The Eagles arrived waving their flag and singing the somewhat menacing theme tune to the movie *Dragnet*. Almost as soon as the game was underway, the Eagles started to chant, "Our pitcher is better than yours," causing the Rattlers to engage in some name-calling. Enraged by the labels "Fatty" and "Tubby," the Eagles responded by getting out some matches and setting fire to the Rattlers' flag. Understandably upset by the demise of their flag, the Rattlers decided to head back to their cabin and plot a raid on the Eagles' camp.

At 10:30 that night the Rattlers put dark paint on their faces and arms and carried out a commando-style attack on the Eagles' cabin. Minutes later, the Eagles awoke to the sound of their beds being turned over and mosquito netting being ripped from their windows. Angry, the Eagles decided to mount a counterattack later that night, but the experimenters banned the assault when they saw the Eagles preparing to use rocks as weapons. Ever resourceful, the Eagles agreed to cancel the evening attack and instead carried out their retaliatory raid the following morning. Arming themselves with sticks and bats, they ransacked the Rattlers' cabin, and then returned to their own cabin and filled their socks with rocks in preparation for a possible return raid.

Within days, the previously peaceful camps had come to resemble a scene from William Golding's novel *Lord of the Flies*. Sherif's

THE AS IF PRINCIPLE

careful selection process had ruled out the likelihood that the boys had any psychopathic tendencies. However, he later noted that if someone had observed the two groups at this point in the experiment, they would have concluded that the boys "were wicked, disturbed, and a vicious bunch of youngsters."

How had this dramatic transformation come to pass? Sherif's experiment was designed to discover whether certain situations possess the power to make level-headed and normal people behave in a highly aggressive way. Before starting the study, Sherif had examined the atrocities that the Greek army had carried out during its invasion of Turkey and concluded that much of the aggression arose from each community having a strong sense of identity and competing for limited resources. To test his hypothesis, Sherif had created a small-scale version of the same situation during his experiment by getting the boys to bond together and then compete over access to the baseball field. All hell was let loose, with the trouble quickly escalating in a series of tit-for-tat responses. Sherif argued that regardless of whether a dispute is over land, power, money, or jobs, the same process quickly turns one group against another.

Disturbed by the level of experimentally induced aggression, Sherif decided to move into the final phase of the study: the experimental production of bonding.

At the start of this phase, the experimenters asked the boys to describe specific members of their own and the opposing group. The boys tended to see the people in their own group as brave and tough and those in the other group as sneaky and untrustworthy.

Sherif first wanted to discover whether it was possible to change the boys' beliefs about one another by bombarding them with information. He asked both groups to attend Sunday religious services and had the minister give pointed appeals for forgiveness, cooperation, and brotherly love.

The boys quietly left the services and within minutes were plotting their attacks on one another.

With the information campaign a failure, Sherif adopted another approach and explored the impact of getting them to help one another.

When people feel that they have bonded with one another, they often act in unison. People who share a religious belief come together to pray, troops march around in step, sports fans cheer for their team, and those attending political rallies applaud speeches. But can behaving as if you are part of such a group help to bond people together?

To find out, Sherif staged a series of fake emergencies that encouraged both the Rattlers and Eagles to stand shoulder to shoulder. In one instance, both groups were told that some vandals had damaged the water supply and that they needed to work together to solve the problem. In reality, there were no vandals and the experimenters had caused the damage themselves by placing two large boulders on one part of the supply. The groups realized that they would both benefit from having drinking water and so worked together to remove the boulders.

In another instance, a "camp attendant" (actually a member of the research team) offered to drive to the nearest town and get both groups some special food. However, his truck suddenly developed a problem, and the Eagles and Rattlers needed to work together to help start the truck.

The results were dramatic. Within just a few days, much of the animosity between the Eagles and the Rattlers vanished into thin air, and the two groups started to bond. On the final night of the project, one of the Eagles took out his ukulele and played a song to entertain the Rattlers. In return, one of the Rattlers performed his Donald Duck imitation, with Sherif's research notes revealing that "this performance was received with great enthusiasm."

The findings from the final phase of Sherif's remarkable experiment demonstrated how the As If principle has the power to change people's beliefs about one another for the better. By getting the Rattlers and Eagles to collaborate, Sherif had managed to get both groups to see each other in a far more positive light.

Inspired by the success of these types of studies, one researcher began to explore whether the same effect could be used in the real world to bring children closer together.

Altogether Now

Want to get a group to quickly bond together and believe in a single cause? Then get them to act in unison.

A few years ago, Scott Wiltermuth and Chip Heath from Stanford University gathered together groups of three students.[37] Some of the groups were asked to walk around the campus normally, while others were formed into a mini-army and asked to march around the same route in step. In another part of the study, groups were asked to listen to a national anthem, and others were asked to sing along and move in time to the music. The people in each of the groups were then asked to play a board game in which they could choose to either help or hinder one another. Those who had been walking in synchrony and singing in unison quickly bonded together and were significantly more likely to help one another during the game.

People who have bonded together often act in unison. Similarly, acting in unison helps people bond together.

In the early 1970s, psychologist Elliot Aronson from the University of Texas was contacted by a superintendent from a local school. The superintendent explained that many of the schools in Austin had recently been desegregated, and so children from a variety of racial backgrounds were sharing a classroom for the first time. Unfortunately, the deep-seated feelings of suspicion and distrust between the different racial groups had resulted in a hostile atmosphere and even violence.

The superintendent asked Aronson whether there was anything he could do to help solve the problem. Aronson visited a few schools and noticed that most of them were producing a strong sense of competition among the students. In the same way that Sherif had created conflict by getting the Rattlers and the Eagles to argue over ownership of the baseball field, so teachers were inadvertently encouraging their classes to compete against one other for good grades. Mindful of the success that Sherif had had by getting his guinea pigs to work together, Aronson invented a new type of learning that has come to be known as the jigsaw method.

Let's imagine that a teacher wants her class to learn about the life and ideas of Martin Luther King Jr. The teacher first divides the students into small groups of five or six pupils, ensuring that each group consists of a diverse mix of gender, race, and ability. Next, the teacher divides the elements of the lesson into several different parts. In the case of Martin Luther King, this might, for example, include information about his childhood, the influence of other leaders on his life, his early protests, his rise to power, his assassination, and his legacy.

One student in each group is then asked to learn about just one of these elements. After they have spent some time finding out what they can, the pupils rearrange themselves so that all of the pupils who have learned about one particular element sit together. Each of these new groups then discusses what they have learned. For example, one group of pupils might share their knowledge about Martin Luther King's early life, while another group might discuss his legacy. At the end of these discussions, the students go back to their original groups, and each pupil presents what he or she has discovered to the group. At the end of the lesson, the teacher gives a short quiz on the material so that the pupils have a chance to discover what they have and haven't learned.

Aronson introduced the jigsaw method into several randomly selected classrooms. Although the pupils in these classrooms spent only a small amount of time using the method, they quickly became far less prejudiced and significantly more self-confident. Not only that, but the children using the method showed a drop in absentee-ism and performed better in their year-end exams.

In his seminal book on social psychology, *The Social Animal,* Aronson discusses the impact of the jigsaw method on a Mexican American pupil named Carlos. At the time of the study, Carlos did not speak English well, and years of education in a substandard, racially segregated school had left him feeling shy and insecure. When Carlos used the jigsaw method, he was required to speak to his group. As he stammered his way through the material, his fellow pupils quickly started to ridicule him. When one of Aronson's researchers heard the exchange, she focused the group's attention on the need to cooper-ate, pointing out that it was vital that they help Carlos speak to the group if they all wanted to do well in the upcoming exam. After a few weeks, Carlos's group became skilled interviewers who were able to ask helpful questions and elicit clear answers. In short, they were behaving as if they liked Carlos, and he quickly became one of the group. As a result, Carlos's self-esteem and performance improved.

Many years later, Carlos came across Aronson's book and rec-ognized himself in it. By that time he had just gained admission to Harvard Law School. Carlos reminisced about Aronson visiting the school ("you were very tall . . . had a big black beard and you were funny and made us all laugh") and how the jigsaw method had con-verted enemies into friends. In his final paragraph Carlos explained why he was writing to Aronson:

> My mother tells me that when I was born I almost died. I was born at home and the cord was wrapped around my neck and the midwife gave me mouth-to-mouth and saved my life. If she was still alive, I would write to her too, to tell her that I grew up smart and good and I'm going to law school. But she died a few years ago. I'm writing to you because, no less than her, you saved my life too.*

---

* www.jigsaw.org/carlos.htm

Chapter 6

Creating a New You

Where we learn how to feel more confident, change our personality, and slow the effects of aging

"No man, for any considerable period, can wear one face to himself and another to the multitude, without finally getting bewildered as to which one is true."

—Nathaniel Hawthorne

I. The Problem with Personality
II. How to Be More Confident
III. Why Clothes Maketh the Man
IV. The New You
V. Turning Back Time

I. The Problem with Personality

Imagine that you go to a job interview and are asked to distill the essence of your personality into just three adjectives. What would you say? Would you, for instance, describe yourself as outgoing or shy? Creative or down-to-earth? A go-getter or laid back? If the interviewer asked you what has made you the person you are today, how would you reply? Do you believe, for example, that your personality is due to genetics, childhood experiences, or events that have happened during your adult life?

Many of the world's greatest thinkers have wrestled with these issues. The Victorian scientist Sir Francis Galton was convinced that character could be best determined by carefully studying the bumps on your skull and the shape of your nose. Sigmund Freud thought that Galton's approach was deeply strange and instead argued that personality is based on the bodily orifice from which you derived greatest pleasure as a child (thus giving rise to the idea of "oral" and "anal" people). Psychiatrist Carl Jung believed that both Galton and Freud were badly mistaken, and instead thought that your identity was determined by the position of the stars when you were born (then again, Jung was a Leo and so was predisposed to come up with silly ideas).

Perhaps not surprisingly, most modern psychologists would not categorize you on the basis of the bumps on your head, preferred bodily orifice, or star sign. Instead, they would see you in terms of your key personality traits.[1]

A few thousand years ago, the eminent Greek philosopher Hippocrates had two interesting thoughts. First, he suggested that all physicians should take the so-called Hippocratic oath and pledge to always act in the best interests of their patients (unless the money was really good). Second, he speculated that differing amounts of blood, phlegm, black bile, and yellow bile caused people to develop one of four personalities: melancholy (the anxious introvert), phlegmatic (the relaxed introvert), sanguine (the relaxed extrovert), and choleric (the anxious extrovert). Although Hippocrates' thoughts about bodily fluids quickly fell out of favor, the idea of trying to fit the apparent complexity of personality into a simple structure has stood the test of time.

In the 1930s Harvard psychologist Gordon Allport came across Hippocrates' work and wondered whether science could help uncover the structure of personality. Allport laboriously worked his way through a large dictionary and noted all of the adjectives that could be used to describe a person's personality. After identifying around four thousand words, Allport (studious, hard working, bored) had had enough and handed the work over to a colleague, Raymond Cattell. Cattell (compassionate, caring, empathic) carefully went down Allport's immense list and eliminated all of the words that described the same general characteristic. Eventually Cattell (compassionate) ended up with a list of about 170 core adjectives.

Several research teams then asked thousands of people to rate themselves with these adjectives and used a sophisticated statistical technique known as factor analysis (don't ask) to analyze the structure in the data. The results suggested that Hippocrates was wrong to think that everyone's personality falls into one of four boxes. Instead, a small number of personality dimensions appear to exist, and everyone falls somewhere along each of them. For example, rather than everyone being either extroverted or introverted, there exists a sliding scale, with, "Yippeee, it's time to party!" at one end and, "Oh, Christ, I'd rather stay home with a good book," at the other. Each of these fundamental dimensions of personality is referred to as a trait.

For the next fifty years or so, psychologists repeatedly locked horns about exactly how many traits were needed to fully describe a person's personality. Cattell, for instance, was convinced that there are sixteen core traits, while the British psychologist Hans Eysenck opted for just three. In the early 1990s, the majority of researchers agreed to split the difference and settle on the existence of five fundamental dimensions: openness (the need for new and unusual experiences), conscientiousness (the tendency for organization and self-discipline), extroversion (the need for stimulation from the outside world and other people), agreeableness (the tendency to care about others), neuroticism (the tendency for emotional instability), and innumeracy (the tendency to struggle with even basic arithmetic).*

Many researchers believe that scores on each of the five personality traits are due in part to genetic makeup. Consider the introvert–extrovert dimension. According to conventional personality theory, your DNA has created a brain that has a certain preset level of arousal, in the same way that your television set may have a preset volume when you first turn it on. If you are an introvert, your brain is naturally aroused, and so you will try to avoid situations that further excite your already stimulated brain. As a result, no matter what situation you are in, you will tend to avoid bright lights and noisy groups of people, and instead gravitate toward much more tranquil activities such as reading and quiet conversation. If you are more of an extrovert, your brain will have a much lower preset level of arousal, and so you will have a need for continuous stimulation. Because of this, no matter what situation you encounter, you will be attracted to the stimulating effects of large groups of people, risk taking, and impulsive behavior.

According to this point of view, your personality is hardwired into your brain, causes you to behave in the same way in many different situations, and doesn't change during your lifetime. Although all of this may sound reasonable, it is far from the full story.

---

* Just kidding about the final one.

Psychologists have frequently tested the notion that people's personalities cause them to exhibit consistently the same behavior patterns in a variety of situations. In one study, for example, counselors working at a summer camp for teenage boys were asked to note down in confidence the degree to which the boys displayed various forms of extroverted behavior, such as talking during mealtimes, seeking the limelight, and initiating conversations.[2] The researchers then carefully analyzed the data by comparing the boys' level of extroversion on odd and even days. The personality-causes-behavior theory predicts that there would be a high level of consistency in the boys' actions, with the extroverted teenagers constantly chatting away and the introverted ones repeatedly hiding away in the corner. In fact, the results failed to show any evidence of such consistency. On one day, one of the boys would be full of beans and very chatty, while on the next day the very same boy would be quiet and withdrawn.

In another experiment, psychologists visited several schools and set up highly realistic situations in order to test the honesty of schoolchildren.[3] The research team presented the children with an opportunity to steal some money that had been "left" on a table, then lie about stealing it to avoid trouble, and finally gave them the chance to change their exam scores. Each time the children's behavior was secretly recorded and compared in the different situations. The personality-causes-behavior theory would predict that the same children tend to steal, lie, and cheat, but the results failed to show such consistency, with a child behaving dishonestly in one situation but then getting in touch with her inner angel in another.

Disillusioned with the notion that personality caused behavior, a small number of researchers started to develop a completely different view of human identity.

In previous chapters, I have presented a large amount of research showing how your behavior can create your emotions, thoughts, and willpower. Smile, and you feel happier. Hold hands with another person, and you find him or her strangely attractive. Tense your muscles, and you develop greater self-control. Inspired by this work, some researchers wondered whether the same process might also explain the relationship between behavior and identity. Rather than your personality causing you to behave in a particular way, could your behavior cause you to develop a particular personality?

Common sense suggests that the chain of causation is:
Extroverted personality— Outgoing behavior
The As If principle suggests that the reality is:
Outgoing behavior— Extroverted personality

If true, this topsy-turvy approach to personality opens up the possibility that you can alter your sense of identity at will. Just by changing the way you behave, you could, for example, quickly become less aggressive, especially likable, and far more confident.

For the past forty years, researchers have examined whether the As If principle can indeed make you feel like a completely new person. Our journey into this real-world version of *Pygmalion* starts with an unusual experiment involving a set of weights and several earthworms.

II. How to Be More Confident

Do you trust your own judgment even when others are questioning your decisions? Are you able to put your mistakes behind you and not spend too much time worrying about what will happen in the future? Do you think that you will do well in most situations? If you have just answered yes to all of these questions, then you probably have fairly high self-esteem, whereas a series of no's suggests that you are a tad insecure.

According to the conventional personality-causes-behavior theory, very low self-esteem has several disadvantages, including encouraging people to endure humiliating and demeaning experiences. However, the As If principle turns this idea on its head. Rather than low self-esteem causing people to put up with demeaning experiences, it suggests that it is the taking part in demeaning experiences that causes people to develop low self-esteem. Psychologist James Laird set out to discover if this was indeed the case.[4]

In Chapter 1, I described how Laird had conducted the first test of the As If principle and discovered that smiling makes people feel happy. Excited by the positive findings from this initial work, Laird has devoted most of his academic career to exploring the power of the principle.

Imagine that you have signed up to participate in one of Laird's studies. You are invited to the laboratory and asked to complete a self-esteem questionnaire. The experimenter then takes you into another room and asks you to sit down at a small table. On the table are some cooking weights, a knife and fork, and a live worm. The experimenter explains that you will have to take part in one of two tasks. One will involve your lifting each of the weights and ordering them according to heaviness. The other involves your cutting up and eating the worm.

The researcher then flips a coin and explains that unfortunately you have been assigned to the worm-eating task. You sit down in front of the wriggling worm and look at the wee beast for a few moments. The experimenter then asks you to complete a second self-esteem questionnaire before preparing the worm.

This experiment was carefully designed to discover whether the As If principle applies to self-esteem. Laird considered that if people saw themselves about to carry out a demeaning and humiliating experience (that is, behaving as if they had low self-esteem), they would be likely to conclude that they did indeed have low self-esteem. As predicted, the self-esteem of the worm eaters crashed and burned. In the same way that smiling caused people to be happy, so taking part in a demeaning experience caused them to develop low self-esteem.

But the study did not end there. Imagine that you have finished the second questionnaire, and just as you pick up the knife and fork, the experimenter rushes across the room and explains he has made a terrible mistake: You should have been given a choice about which task to complete. Do you choose to continue to consume the worm or switch to the weights?

Laird knew that people with low self-esteem often believe that they deserve bad experiences, and he wanted to discover whether the laboratory-induced low self-esteem would alter people's behavior. None of the volunteers who were randomly assigned to the weights switched to the worm eating. However, and remarkably, only 20 percent of the people assigned to eating the worm switched to the weights. Although these people now had the option to switch to a less awful task, their experimentally created low self-esteem caused the vast majority of them to choose to eat the worm (although just as the participants were about to munch the worm, the experimenters rushed across the room and stopped the study).

When Laird's study was published, some psychologists criticized the method, arguing that perhaps the participants were merely role playing, safe in the knowledge that the researchers would never allow them to actually eat a worm. As a result, other researchers repeated the study, this time using large, edible caterpillars.[5] In this second study, the participants actually consumed the caterpillars, and the results replicated Laird's original findings.

This effect helps explain why people who have had the ill fortune to experience a random but negative life event often end up with low self-esteem and even blame themselves for it. People who have been the victims of random violent assaults often feel that they somehow caused the attack, and those suffering from a terminal illness wonder what they have done to deserve their fate. As predicted by the As If principle, their sense of identity is the direct and unfortunate result of having been forced to endure an unpleasant event.

Unfortunately, once the process has started, it feeds on itself. Those low in self-esteem endure more negative events, which in turn cause an even greater decline in their esteem. The good news is that the same idea can be used to quickly boost self-esteem and confidence.

Most courses designed to improve self-esteem are based on the notion that low self-esteem and poor confidence are a result of the way people think about themselves, and therefore encourage participants to focus on instances in their life when they have done well or ask them to visualize themselves being more assertive. In contrast, the As If principle suggests that it would be much quicker and far more effective to ask people to change their behavior.

In one early study, experimenters assembled a group of participants to apparently examine whether a newly developed pair of plastic glasses might affect people's perception.[6] The participants were split into two groups, and both groups were given the same intelligence and personality tests. Half were asked to complete the tests with no special conditions, while the others were given a pair of glasses with clear lenses. Because people associate glasses with intelligence, the experimenters speculated that merely wearing them would make people suddenly feel more confident and clever. They were right. The actual scores on the intelligence tests didn't differ between the two groups, but when they were wearing the glasses, the participants rated themselves as more stable, competent, and scholarly.

Then there is the issue of posing. Researcher Dana Carney from Columbia University knew that confident people tend to feel good about themselves, take more risks, and have higher levels of testosterone (a chemical associated with dominance) and lower levels of cortisol (a chemical associated with stress). What would happen, wondered Carney, if a group of people were asked to behave in a dominant way? To find out, she and her colleagues gathered together a group of participants, told them that they were there to help assess a new heart monitoring system, and split them into two groups.[7]

The participants in one group were placed in one of two power poses (see Figures 4 and 5). Some were seated at a desk, asked to put their feet up on the table, look up, and interlock their hands behind the back of their head. Others were asked to stand behind a desk and lean forward, placing both hands palm down on the desk. Those in the other group adopted one of two poses that weren't associated with dominance (see Figures 6 and 7). Some of these participants were asked to place their feet on the floor, with hands in their laps, and look at the ground. Others stood up, and crossed their arms and legs.

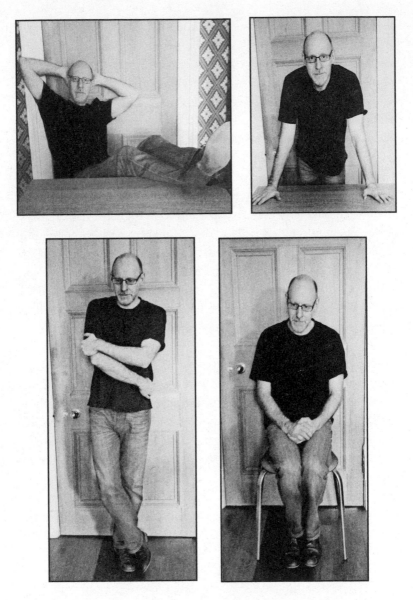

After just one minute of posing, all participants were asked to rate how "powerful" and "in charge" they felt. The posing had a significant effect on their self-esteem, with those who had been placed in the power poses giving higher ratings than the others. But this was just the tip of the iceberg.

Participants were then given a quick risk-taking test. They were handed two dollars and told that they could either keep the money or gamble it on the flip of a coin. If they won the toss, they would double their money and get four dollars; if they lost the toss, they would go home with nothing. In line with the "power posing makes you more of a risk taker" hypothesis, more than 80 percent of those put in the dominant position took the gamble compared with just 60 percent of others.

In the final part of the study, the researchers turned their attention to the chemicals coursing through the volunteers' veins. Both before and after adopting the experimental posing, the researchers asked all of the participants to chew on gum for a few minutes to build up a ready supply of saliva (noting that this technique was just as good as "passive drool procedures") and then spit into a test tube. After analyzing the contents of the tube, the researchers could see that compared to those sitting with their hands in their laps, those who were power posing had significantly higher levels of testosterone and lower levels of cortisol than before the test. In short, just a minute behaving as if they were more dominant had changed the chemical makeup of their bodies.

Finally, if you haven't got time to strike a powerful pose, just make a fist.[8] Psychologist Thomas Schubert asked a group of men to rate how confident they felt, then to form their hand into a fist for a few seconds (under the guise of playing "rock-scissors-paper"), and then to rerate their confidence. The participants' bodies influenced their brains, with the men enjoying a significant boost in confidence because they had spent a few moments forming a fist.

The Confidence Trick

This exercise requires a pen, a piece of paper, and two hands.

1. Rate how confident you feel about yourself on a scale between 1 (not at all confident) and 7 (very confident).

2. Take a look at this list of adjectives and choose three that reflect your best traits and three that reflect your worst qualities:

loyal, affectionate, aloof, ambitious, unmotivated, secretive, caring, callous, cheerful, grumpy, considerate, thoughtless, cooperative, unhelpful, courageous, rude, indecisive, enthusiastic, apathetic, flexible, stubborn, unforgiving, focused, frugal, generous, grateful, hard working, lazy, honest, dishonest, humble, arrogant, jealous, immature, modest, optimistic, pessimistic, punctual, self-confident, insecure, sincere, disorganized, pretentious, ostentatious

3. Place the pen in your nondominant hand and slowly write the three negative traits that you selected on the paper.

4. Place the pen in your dominant hand and slowly write the three positive traits you selected on the piece of paper.

5. Rate how confident you feel about yourself on a scale between 1 (not at all confident) and 7 (very confident).

Did the exercise make you feel more confident?

This exercise is based on work conducted by psychologists Pablo Briñol from the Universidad Autónoma de Madrid.[9] Briñol told participants that they were taking part in a study about graphology and asked them to write down their best or worst qualities using either their dominant or nondominant hand. Immediately afterward, all of the participants rated their self-esteem and confidence.

The experimenters knew that when the participants used their nondominant hand, they would see themselves produce very shaky handwriting, and so were behaving as if they didn't have much confidence in their words.

As a result, the researchers predicted that listing positive traits with their nondominant hand would decrease the participants' self-esteem, whereas writing their negative traits with their nondominant hand would make them feel more confident and motivated. This is exactly what their results revealed.

If you want a quick and effective way of increasing your confidence, the message is clear: use the write stuff.

III. Why Clothes Maketh the Man

John Howard Griffin lived a remarkable life. Born in Texas in 1920, Griffin traveled to Europe at a young age and trained as a musicologist specializing in Gregorian chanting. When World War II broke out, he worked for the French Resistance and helped smuggle Austrian Jews to safety. After the war, he returned to America, became an investigative journalist, and decided to highlight the plight of African Americans living in the South.

However, rather than simply write about racism, he decided to carry out an extraordinary experiment in order to experience it firsthand. Working closely with a skilled dermatologist, Griffin used a mixture of artificial pigments, drugs, and sun lamp treatments to darken his white skin. Once the transformation was complete, he shaved off all his hair and so, to the casual observer, he appeared to be an African American. He then toured around several southern states by bus and hitchhiking, experiencing the hatred and discrimination genuine African Americans faced every day.

At the start of his bestselling book about the project,[10] Griffin describes how he looked at himself in a mirror once the transformation was complete and presents a vivid account of the impact that this reflection had on his sense of self:

> I had expected to see myself disguised, but this was something else. I was imprisoned in the flesh of an utter stranger, an unsympathetic one with whom I felt no kinship. . . . I looked into the mirror and saw reflected nothing of the white John Griffin's past. No, the reflections

led back to Africa, back to the shanty and the ghetto, back to the fruit-less struggles against the mark of blackness. . . . I had tampered with the mystery of existence and I had lost the sense of my own being. This is what devastated me. The Griffin that was had become invisible.

By changing the color of his skin, Griffin felt like a different per-son. Throughout his life, Griffin would have looked in mirrors and seen a Caucasian man. Assuming that skin color was an important part of his self-identity, he would have seen himself as possessing the background and traits associated with his appearance. After his dra-matic transformation, he saw himself resembling an African Ameri-can and unconsciously used this image to construct a new sense of identity. Within seconds, he experienced his old self crumbling away and a new identity forming.

Most people are unlikely to follow in Griffin's footsteps and try to change their skin color. However, the same principle applies to some-thing that is much easier to alter: your clothing. We frequently judge other people on the basis of the clothes they are wearing. If you see a man in an expensive suit, you tend to automatically assume that he is successful and competent. See the same person dressed in a kaftan or flowery shirt, and you might assume that he is a creative type. Bump into the man wearing large shoes, a red nose, and a clown's pants, and you know it is time to run.

These differences in perception often cause us to behave in very different ways. For example, Nicolas Guéguen from the Université de Bretagne Sud dressed up men in either civilian clothing or a fire-man's outfit and had them approach more than two hundred ran-domly chosen women on the street.[11] Each time the man caught the woman's eye, he came out with the same prepared script:

> Hello. My name's Antoine. I just want to say that I think you're really pretty. I have to go to work this afternoon but I wonder if you would give me your phone number. I'll phone you later and we can have a drink together.

Guéguen carefully analyzed the percentage of women who were happy to give their telephone number and discovered that the uniform had a remarkable effect. When the men were dressed in civilian clothing, just 8 percent of the women gave their telephone number. However, when exactly the same men were dressed as firefighters, they enjoyed a 22 percent success rate.

In a similar study, John Marshall Townsend from Syracuse University dressed the same group of actors in either a Burger King uniform or a fashionable suit, showed photographs of the actors to women, and asked them to rate whether they would be willing to have sex with them.[12] Clothes madeth the man, with more women indicating that they were willing to sleep with the man when he was wearing a suit as opposed to a Burger King uniform.

Other research shows that even the smallest changes can have a big effect. In another study, a psychologist posing as a market researcher approached various people and asked whether they would mind taking part in a survey.[13] Half the time the man wore a tie, and the rest of the time he did not. This small difference had a huge effect, with more than 90 percent of people agreeing to carry out the survey when he was sporting a tie compared with just 30 percent when he wasn't.

In the same way that the clothes people wear clearly affects our perception of them, do the clothes that you wear influence the way in which you see yourself? Most proponents of the personality-causes-behavior theory would argue that your sense of self has slowly developed over a period of years and isn't going to be influenced by something as transitory as putting on a new shirt or changing your shoes. In contrast, the As If principle predicts that dressing as if you are a certain type of person will affect your sense of identity. To discover which theory was correct, Mark Frank from Cornell University conducted a series of studies.[14]

Frank knew that people tended to associate black clothing with authoritarian and aggressive behavior and wondered whether simply wearing such clothing might change the way in which people behaved. Luckily for him, the data needed to test his hypothesis were already available. Frank searched through the records of the National Football League, comparing the data from teams that wore black outfits with others. Frank identified the five teams that wore black outfits, including the Los Angeles Raiders, Pittsburgh Steelers, and Cincinnati Bengals, and started to look at their behavior on the field.

In football, infringements can be punished, with the offending team being moved back by five, ten, or fifteen yards. Frank calculated the average number of yards each team was moved back during each game and discovered a remarkable pattern: the teams dressed in black were moved back significantly more than the others, suggesting that they tended to engage in especially aggressive behavior.

Encouraged by these early results, Frank moved on to the records from the National Hockey League and again compared teams wearing black uniforms with others. In ice hockey, violations can result in players having to sit out for two, five, or ten minutes depending on the severity of the violation. Frank discovered that those in black spent significantly more time on the bench.

The hockey data also allowed Frank to conduct an especially clever test of his hypothesis because two of the teams, the Pittsburgh Penguins and the Vancouver Canucks, switched from nonblack to black uniforms. The "wear black, become aggressive" effect emerged. Before switching to black, the players from both teams rarely frequented the benches. Afterward they became a near-permanent fixture.

Most researchers would have left it there. However, Frank knew that other researchers would be skeptical of his idea and perhaps argue that aggressive players were attracted to teams with black outfits. The only way to settle the matter was to conduct an experiment. He assembled a group of willing volunteers and randomly split them into two groups. One of the groups was dressed in black and the other in white. Both groups were then told that they were going to be split into small teams and asked to play a variety of games. The experimenters presented the participants with a list of games and asked them which ones they would like to play. Unknown to the participants, the games had been specially chosen to vary in aggressiveness. Some of the games, like the dart gun game, were highly aggressive, while others, such as the putting contest, were far more passive. The people who had dressed in black chose far more aggressive games than those dressed in white.

Other research suggests that this effect is far from a black-and-white issue.

A study conducted by Robert Johnson from Arkansas State University assembled a group of participants and explained that they were going to be offered an opportunity to give another person electric shocks.[15] The experimenter described how each participant was going to be photographed before giving the shocks, but it was important that the clothing was covered up in the photograph. And how was this anonymity going to be achieved? The helpful experimenter had brought along two types of outfit. Half of the participants were asked to wear a robe resembling a Ku Klux Klan outfit, with the experimenter justifying the clothing by mumbling, "I'm not much of a seamstress; this thing came out looking kind of Ku Klux Klannish." In contrast, the other participants were given outfits that made them look like nurses ("I was fortunate the hospital recovery room let me borrow these nurses' gowns to use in the study").

In the next stage, the experimenter said that there was a person in the next room trying to learn a list of words and asked the participants to administer an electric shock to the person whenever he or she made a mistake. In reality, the person in the next room was a stooge, and the electric shock apparatus was entirely fake. Whenever the participant heard the person in the next room make a mistake, he or she could choose to increase or decrease the alleged shock. Participants dressed in outfits resembling those of the Ku Klux Klan chose to deliver significantly higher shocks than those wearing the nurse outfits, exactly as the As If principle predicted.

The same effects have been observed beyond the laboratory.

In 1969, the police in Menlo Park, California, decided to try to improve police community relations by getting rid of their navy blue paramilitary-style uniforms and changing into a more relaxed look.[16] The officers were asked to wear a forest green sports blazer, black slacks, a white shirt, and a black tie. They were also asked to conceal their weapon under the blazer. Soon word spread, and more than four hundred other American police departments experimented with the same informal uniform. Eighteen months into the experiment, researchers had the officers complete various tests, with the results showing that when stripped of their symbols of authority, the officers gradually adopted a new role of police officer as public service officer. In line with this new identity, the police officers displayed fewer authoritarian characteristics compared with their colleagues in more formal attire. During the same period, injuries to civilians by the police dropped by 50 percent.

The message is clear: the way you dress directly influences who you think you are. Don a black shirt, and you start to become an authoritarian and aggressive person. Slip into something far more comfortable, and you turn into a more tolerant and giving individual. For years, psychologists have urged job candidates to dress in a business-like outfit before an important interview in the belief that this clothing will have a positive impact on the interviewer. The As If principle suggests that such attire will also have a profound and perhaps even more important effect on candidates. By dressing well, they will see themselves as more successful, and this will help them perform far better in the interview. Clothes do not just maketh the man. They maketh every man, woman, and child.

Thinking Outside the Box

Want to instantly become a more creative person? Then try this two-part experiment.

First, think of as many uses for a pencil as possible. For example, you could use a pencil as a magic wand or dowel. However, before you start writing down your ideas, spend sixty seconds walking around your room, ensuring that your path forms a rectangle or box (that is, walk in straight lines and perform ninety-degree turns).

Now get a piece of letter-sized paper and spend the next sixty seconds jotting down your alternative uses for the pencil.

For the second part of the experiment, think of as many uses for a piece of paper as possible. For example, you might fold up the paper and use it as a hat or doorstop. However, before you start writing down your ideas, spend sixty seconds walking around your room, but this time ensure that your path is far curvier and more unpredictable than before (that is, avoid walking in straight lines and instead trace out whatever shape you like).

Now spend the next sixty seconds jotting down your alternative uses for the piece of paper.

Research carried out by Angela Leung and her colleagues at the Singapore Management University suggests that your behavior directly influences your level of creativity.[17] In one experiment, the researchers asked some participants to sit inside a five-foot-square box, while others sat outside the box. In another study, some participants walked around a room in straight lines, while others walked along far more random and curvy paths. After completing the exercises, all the participants were asked to carry out various creativity tasks. Those who had literally been thinking outside the box and walking in a more free-flowing way obtained significantly higher creativity scores than the other participants. Behaving in a creative way directly influenced the way the participants thought.

According to these findings, you should have found it much easier to produce alternative uses for the paper than the pencil. Want to get your creative juices flowing? Then forget about expensive lateral-thinking courses and go for a long, unpredictable, and meandering walk.

For an extra boost, try acting as if you are creative. Take a sheet of blank paper and spend a few moments thinking about how you can transform it into a work of art. Before deciding on your final course of action, look at this list and see if anything appeals to you:

Use the paper in a creative way by

... cutting out a silhouette of a skyline or person
... folding it into a box or model building
... making a random squiggle on the paper and then converting this squiggle into a picture
... making a sculpture
... using it to add a pop-up element to this book
... using it to create an interesting shadow
... turning it into a copy of a famous painting or work of art
... turning it into a motivational poster
... making an origami figure, such as a frog, bird, paper airplane, or swan
... making a picture by simply creasing the paper
... folding it up and ripping out pieces to make a snowflake design
... folding it up and tearing parts away to make a chain of people
... making a flip book
... pleating the paper to create your own accordion
... tearing it into tiny pieces and rearranging them to create a work of art

... using it to make an item of clothing or jewelry (such as a hat, ring, or badge)

... making a rubbing of a surface

... creating your own paper currency for a magical kingdom

... making two holes in it and wearing it as a mask

IV. The New You

By law, any contract for a book about the history of social psychology has to contain Clause 4.6.8.3.2. This clause explicitly states that at some point in the book, the author is legally obliged to describe the Zimbardo prison experiment. As a result of this clause, many authors are forced to crow-bar in the study, often placing it between a lurid description of Milgram's infamous electric shock study and a closing paragraph about the apparent banality of evil. Fortunately, I am not faced with a similar dilemma because Zimbardo's classic experiment plays a central role in our journey into the surprising relationship between the As If principle and identity.

Philip Zimbardo was born into poverty in New York City's South Bronx ghetto during the Depression. Fascinated by the impact that people's surroundings had on their behavior, Zimbardo studied psychology throughout the 1960s and eventually joined the faculty at Stanford University, where he conducted his now infamous experiment.[18]

Prior to starting the study, he converted the basement rooms of Stanford's Psychology Department into a mock jail. Several of the smaller rooms were turned into cells by replacing their doors with steel bars. Other areas were converted into guards' living quarters and a prison "yard." The mock prison also contained several two-way mirrors and hidden cameras that allowed the experimenters to observe and record the participants' behavior.

Zimbardo then placed an advertisement in a local newspaper appealing for men to take part in a two-week study examining prison

211

life. All applicants were sent an extensive questionnaire about their background, psychological health, and any prior history of criminal activity. Zimbardo carefully examined the replies and invited the twenty-four applicants who showed the highest levels of psychological stability and lowest levels of previous antisocial activity to take part. He randomly assigned half of them to play the role of the "prisoners" and the other half to be the "guards." Zimbardo himself decided to adopt the role of the prison superintendent (a move that he later described as "a serious error in judgment").

Just prior to the study, there had been violent clashes between police officers and antiwar protesters on the Stanford campus, and Zimbardo found out that the city police chief was eager to improve his relationship with the university. He asked the chief whether he might be able to assign some officers to help with the initial part of the study and the chief agreed. On the first morning of the experiment, nine of the "prisoners" were unexpectedly arrested in their homes by the Palo Alto City Police Department. All of them were charged with suspicion of burglary or armed robbery, handcuffed, and driven to their local police station. The officers then searched and fingerprinted the prisoners, blindfolded them, and drove them to Zimbardo's mock prison.

Meanwhile, the volunteers playing the role of the guards were dressed in khaki uniforms and given a whistle, mirror sunglasses, and a wooden baton. They were asked to look after the prison by working in three-man shifts, with each shift lasting eight hours.

Life was not much fun for the prisoners. When they arrived, the guards gave each of the prisoners an identification number, strip-searched them, took away their clothing, and made them wear an ill-fitting smock. The prisoners were not allowed to wear any underclothes and had a chain attached to one of their ankles. They lived in the prison twenty-four hours a day, had three bland meals a day, and were allowed to visit the bathroom only three times in any twenty-four-hour period.

The guards quickly started to behave in a way that was consistent with their role. They often became highly authoritarian, referring to the prisoners by their identification numbers, and making verbal threats of violence. If the prisoners refused to obey these threats, the guards forced them repeatedly to cite their identification numbers, refused them permission to visit the bathroom, and removed bedding from their cells. On the second day of the experiment, some of the prisoners decided to stage a revolt by blockading their cell doors and ripping off their identification numbers. In response, the guards attacked the prisoners with fire extinguishers (ironically, the University Ethics Committee had insisted on the extinguishers to safeguard the prisoners) and punished those involved by stripping them naked, placing them in solitary confinement, and forcing them to do push-ups.

At the time of the study, Christina Maslach was a psychology postgraduate at Stanford University and involved in a romantic relationship with Zimbardo. Curious to discover how her partner's experiment was progressing, Maslach visited the prison and briefly chatted to one of the off-duty guards. The man came across as friendly and affable. A few moments later, one of the experimenters asked Maslach if she would like to see one of the guards in action. They explained that they had labeled the guard "John Wayne" because of the aggressive way he handled the prisoners. When Maslach looked, she was stunned to see that "John Wayne" was actually the friendly guard she had met earlier. When he wasn't in the prison environment, the guard seemed to be pleasant and placid. Inside the mock prison, he appeared to become a completely different person, shouting at the prisoners and treating them poorly.

After visiting the mock prison, Maslach had a heated argument with Zimbardo. She thought that the situation had spiraled out of control and should be stopped. Normally gentle and sensitive, Zimbardo appeared distant and inclined to let the project continue. Maslach was stunned. She realized that Zimbardo had adopted his role as the prison supervisor and was no longer standing outside the study but had instead become an integral part of it. As the argument continued, Zimbardo realized what had happened and decided to call a halt to the study. Although originally planned to run for two weeks, the experiment was abruptly terminated after only six days.

A key part of Zimbardo's experiment involved exploring whether acting as if they were a prisoner or guard would influence participants' identities. The results were as fast as they were dramatic. After the experiment, one of the guards noted:

> I really thought that I was incapable of this kind of behavior. I was dismayed that I could have acted in a manner that is so absolutely unaccustomed to anything that I would really dream of doing. And while I was doing it I didn't feel any regret or any guilt. It was only afterwards when I began to reflect on what I had done that it began to dawn on me and I realized that this was part of me that I hadn't really noticed before.

Similarly, those playing the role of the prisoners also changed their identities, with most of them becoming extremely passive and far more submissive. These rapid changes often had an extremely negative effect on the prisoners. Doug Korpi (Prisoner 8612) suffered one of the more extreme emotional reactions and was released during day 2 of the study (fascinated by his behavior during the study, Korpi went on to study forensic psychology and work in Californian prisons). A few days later, four other prisoners were released early due to evidence of anxiety, depression, and anger.

Zimbardo's classic study illustrates the power of acting As If. People's sense of unique identity comes from their name, clothing, and appearance. In the prison study, all of these were removed, causing people to lose their own sense of identity and replace it with the role they had been assigned.

Dressing and acting as if they were prisoners or guards, the participants had started to think in ways that they considered consistent with their roles. One group quickly became aggressive and authoritarian, the other passive and conforming.

Zimbardo conducted his study inside a mock prison and used the As If principle to create aggression and anxiety. Other research has shown that the same principle operates in everyday life and applies to many aspects of identity. In one study, for instance, researchers tracked the lives of a group of women over the course of several years and discovered that those who were given extra responsibilities in the workplace developed more assertive personalities.[19] In another study, employees who were given more demanding work became more flexible and confident.[20] A large part of people's personalities are not fixed. Instead, they often accept the roles that they have been assigned by themselves and others, behave accordingly, and then develop an identity that matches this role.

Perhaps most intriguing of all, other psychologists have explored how people can use this effect to transform their lives for the better.

Lights, Camera, Action

George Kelly was born in 1905 on a Kansas farm. After high school, he received a degree in physics, moved to Minnesota, and taught public speaking. Then he stopped teaching public speaking, enrolled at the State University of Iowa, and was eventually awarded a doctorate in psychology. Recognizing the difficulties of farming families during the Depression, Kelly decided to take his psychology on the road and act as a traveling psychotherapist.

At first he adopted a Freudian approach, asking farmers to lie down on his couch and describe their dreams. However, he soon found Freudian theory too esoteric for the down-to-earth farmers and so started to develop more practical ways of tackling their problems.

One of Kelly's first creations has come to be known as mirror time. During these sessions, people would be encouraged to spend thirty minutes sitting in front of a mirror looking at their reflection and thinking about what they saw. Did they like or dislike the person in front of them? What were the differences between the person in the mirror and the person they wanted to be? What did they see in their face that others missed?

Although people often enjoyed staring longingly into their own eyes, Kelly wasn't convinced that these moments of deep reflection were proving especially helpful, and so he decided to draw on his experience of teaching public speaking to encourage people to explore other ways of seeing the world.

His extensive therapeutic experience led him to believe that people's personalities were not fixed, and in the same way that actors play many different roles during their career, so people can change their identities throughout their lifetime. Moreover, Kelly was convinced that the way in which people saw themselves was often at the root of their problems and that effective therapy involved helping clients adopt a less problematic identity. Kelly christened this approach "fixed-role therapy" and over time developed a series of effective techniques for getting people to adopt a new identity.[21]

The first stage of fixed-role therapy usually involves various exercises that are designed to help you understand how you currently see yourself. One of the most popular exercises involves comparing yourself to several other individuals you know in order to identify the core psychological dimensions that you use to classify people. Another exercise might be writing a short description of yourself from someone else's perspective (see the "Who Do You Think You Are?" box).

On the basis of the results, you design a new identity for yourself. This new identity might entail a radical overhaul of your personality or just tweaking a few small aspects. You then spend some time thinking about how this new you would behave in the types of situations you encounter every day and possibly involve some role playing to help solidify these new behavioral patterns.

In the next part of the process, you play your new role for about two weeks. Kelly's research then revealed something strange: after spending a few weeks behaving in a completely different way, many people started to forget that they were playing a role and began to develop a new identity. Many of Kelly's clients reported that the new role seemed as though it had always been their real self and that it was only now that they were becoming fully aware of it.

People had created a new sense of identity by behaving like the person that they wanted to be, exactly as predicted by the As If principle. The same principle can also help bring people together by getting them to see how the world looks to others. In one study, for example, one group of students was asked to behave as if they had recently been paralyzed in a road accident and were now confined to a wheelchair.[22] The students spent twenty-five minutes working their way around a prespecified route in a wheelchair and so had to navigate several lifts, ramps, and doors. Another group of students walked behind the wheelchairs and witnessed everything that happened. Both groups were then asked about their attitudes about issues related to disability, including, for example, whether public funds should be spent on a new rehabilitation center. A remarkable difference emerged between the groups, with those who had spent time in a wheelchair showing far greater empathy for the disabled. The same principle is often used in a form of therapy known as psychodrama. By getting clients to adopt different personas and sometimes even role-play friends and colleagues, they see their life from several very different perspectives.

Kelly's research has provided new and improved identities for tens of thousands of people across the world. New technology has taken the idea to previously unimaginable heights.

Who Do You Think You Are?

PART ONE

Do you want to find out how you see yourself and others? The following two exercises, based on the work of George

Kelly, will give you an insight into how you perceive your personality.*

Exercise One: Your Constructs

This four-stage exercise usually takes about twenty minutes to complete and is designed to give you insight into the core dimensions that you use to see yourself and others. Write all of your answers on a piece of paper.

Stage 1: Think of five people you know well—perhaps your mother, your father, your closest friend, your boss, your partner, a work colleague, or a past lover. Write the names of these people on your piece of paper like this:

Person 1: _____

Person 2: _____

Person 3: _____

Person 4: _____

Person 5: _____

Stage 2: Now look at the first row of the table below. The columns marked "Person 1" and "Person 2" contain Xs. Think of one way in which the personalities of person 1 and person 2 differ from you. Perhaps, for example, person 1 and person 2 are outgoing, and you are shy. Or perhaps they are both somewhat stingy, and you more giving. Copy the table onto your paper and write down the way in which person 1 and person 2 are similar in the column marked "Way in

---

* The exercises described here are designed to provide a general insight into the sorts of techniques that are used by psychologists. If you believe that you have a serious problem in your life, consult a professional.

which they are similar" and the opposite trait (which should describe your personality) in the column marked "Me."

Person 1	Person 2	Person 3	Person 4	Person 5	Way in which they are similar	Me
X	X					
	X	X				
		X	X			
			X	X		
X		X				
	X		X			
		X		X		
X			X			
	X			X		

Stage 3. Move into the next row and repeat the same process. Once again, think about how person 2 and person 3 are similar, and how they differ from you. Carry on like this down the table, trying to come up with different traits each time.

Stage 4. Look at the resulting list of traits in the "Me" column and try to identify some commonalities. Do words like *anxious* and *relaxed* appear a lot? Or perhaps the words *outgoing* and *shy* make regular appearances. These are the core psychological constructs that you use to see yourself and others.

Here is a typical completed table.
Person 1: __ John_____
Person 2: __ Katie _____
Person 3: __ Jenny_____
Person 4: __ David_____
Person 5: __ Erica_____

Person 1	Person 2	Person 3	Person 4	Person 5	Way in which they are similar	Me
X	X				Like detail	Bigger picture thinker
	X	X			Highly artistic	More down to earth
		X	X		Anxious	Relaxed
			X	X	Pessimistic	Optimistic
X		X			Disorganized	Organized
	X		X		Conscientious	Less reliable
		X		X	Agreeable	More hard-headed
X			X		Shy	Outgoing
	X			X	Neurotic	Relaxed

Exercise Two: The Description

PART ONE

Spend about twenty minutes writing a brief description of yourself. Write this description from the third-person perspective, perhaps as you think a close friend or work colleague sees you.

PART TWO

The following stages are designed to help you create and adopt new aspects of your identity.[23]

Stage 1. Take a look at the constructs that you used to describe yourself in Exercise One. Do you think that any of these constructs are negative or problematic? Now look at the description that you produced in Exercise Two. Does that contain any clues about which aspects of your identity you would like to change? Perhaps, for example, you don't think that you are very confident, are struggling to make friends, are overly aggressive, or seem a bit selfish.

Stage 2. Use this information as a basis for the new you. If you are struggling to think about how you want to change, perhaps draw on elements that you admire in friends, colleagues, role models, and even fictional characters in books, movies, and plays. Alternatively, look at the list of character strengths described in the table that follows and choose one or more that appear especially appealing.[24]

Character Strength	Brief Description
Creativity	Good at thinking of novel ways to do things
Curiosity	Having an interest in exploring and discovering
Open-mindedness	Being willing to examine issues from several perspectives
Bravery	Not running away from a threat or challenge
Persistence	Keeping with it when the going gets tough
Vitality	Approaching life with zest and energy
Love	Being able to form close relationships with others
Kindness	Enjoying giving to other people
Citizenship	Being a team player and supporting those around you
Leadership	Taking responsibility and moving forward
Forgiveness	Being able to forgive those who have done wrong
Humility	Not drawing attention to your achievements
Prudence	Expressing self-control and not being overly impulsive
Gratitude	Being thankful for the good things in your life
Hope	Expecting that good things will happen and being prepared to work to achieve them
Humor	Seeing the funny side of life and being light-hearted

Next, write a short description of the new you, focusing on how you will behave differently in everyday situations. For example, let's imagine that you tend to be overly aggressive and frequently get into arguments with your friends and colleagues. The new you might be a lot more relaxed and fun to be with. If that were the case, how would you behave? Would you joke around? Would you ask other people for their thoughts and opinions and accept these comments rather than use them as the basis for an argument? Would you make a special effort to give praise and encouragement?

Alternatively, let's imagine that a few people have remarked on the fact that you are stingy, and you have decided to change this part of your character. Would the new you contribute to charity, give generous gifts to others, and go out of your way to help those around you?[25]

Or maybe you would like to be more confident. If so, do you have an especially assertive friend or colleague who could act as an inspiration for the new you? How does that person behave in the types of situations that you struggle with? Can you pretend to be that person and act differently from the old you in these situations?

Stage 3. Spend around two weeks role-playing your new identity. Focus on changing your behavior rather than trying to change the way you think. To help with this process you might also ask a close friend or family member to role-play a few common situations with the new you. Also, rather than seeing yourself as undergoing a permanent change, it might be helpful to think of your old personality as being on vacation for two weeks, and so you have an opportunity to act like a different person. It is important, however, that you play out your new role twenty-four hours a day, even when you're alone. The As If principle will cause you to feel like a new person, and the new you will soon become part of your actual identity.

Orcs and Elves

Jeremy Bailenson runs the Virtual Human Interaction Lab at Stanford University. Much of his work is creating computer-generated representations of humans (referred to as avatars) and having them move around virtual worlds. Bailenson came across the As If principle and wondered whether it would also operate in the imaginary worlds that he created. Would, for instance, people who were given a tall avatar in a computer game feel more assertive in real life? Or would those who were given an avatar dressed in black become more aggressive?

The possibilities seemed endless, but first Bailenson had to discover whether the principle operated in the virtual world. To find out, he turned to the international phenomenon that is World of Warcraft.

World of Warcraft is a hugely popular online fantasy game in which millions of people across the globe fight against one another in a virtual land. The game involves taking part in "epic sieges" and "a host of legendary experiences," with each player trying to get through sixty levels of increasing difficulty. Before starting the game, players have to create their own avatar. These virtual identities are based on one of eight different "races" (including Gnomes, Night Elves, Orcs, Trolls, and Humans), with each race having a predetermined height (for example, Gnomes are somewhat short, while Trolls are much taller).

Bailenson and his colleague Nick Yee knew that in real life, taller people are more assertive than shorter people and wondered whether the same effect would emerge among the World of Warcraft avatars.[26] To find out, they examined the data from more than seventy-six thousand players, looking at the relationship between the height of each player's avatar and how far the player had progressed in the game. The virtual world mimicked the real world, with players having relatively tall avatars (think Trolls and Orcs) outperforming those with shorter virtual selves (think Dwarfs or Gnomes). These results have two important implications: they suggest that the As If principle functions in virtual worlds, and, on a more practical level, if you want to gain the edge in World of Warcraft, become a Troll rather than a Gnome.

Although Bailenson and Yee were excited by their results, they realized that there were two problems. First, critics might argue that more competent or assertive players had simply chosen taller avatars at the start of the game. Also, even if the height of a person's avatar did change the way this player behaved online, there was no guarantee that it would affect him in the real world. Bailenson and Yee tackled both issues in a second experiment.

In this study, a group of students were asked to wear virtual reality glasses that allowed them to see a virtual version of themselves on a computer screen. To make the situation as realistic as possible, the research team attached high-tech sensors to the students' faces, arms, and legs and ensured that the virtual figure copied their movements. If the student looked to the left, so did the avatar. If the person started to jog, the avatar trotted along too. The close match-up between the participants' actual movements and those of their virtual self resulted in the students' quickly believing they were the figure on the screen.

At the start of the experiment, the researchers randomly assigned a tall or short avatar to each participant. Then, after the students had spent some time in the virtual world, they were asked to remove their headsets, come back to the real world, meet another participant, and play a game called Ultimatum.

In this game, one player has just one shot at proposing a way of dividing a hundred dollars between both players. If the second player accepts the offer, the cash is divided. If, however, the second player rejects the offer, then neither player gets anything. The test has been used in psychology departments across the world for years, with the first player's opening offer being seen as a good reflection of how assertive and aggressive this person is.

The players who had been given a tall avatar made more assertive opening offers than those who had been assigned a short avatar, as predicted by the As If principle. The difference was far from trivial, with those with taller virtual selves offering, on average, a sixty-forty split, while those with shorter avatars offered a fifty-fifty split. Not only that, but those with taller avatars were also more assertive when it came to accepting or rejecting the offers of others. Around 60 percent of those with tall virtual selves rejected a seventy-five twenty-five split in favor of the other player, compared with just 30 percent of those with short avatars.

Bailenson and Yee's work paved the way for an explosion of similar studies. In one study, some volunteers saw their avatars run on treadmills, while another group saw their virtual selves lounging around.[27] When the researchers tracked the participants over time, they discovered that those who had seen themselves running around in the virtual world were far more likely to exercise for real. In another experiment, people who inhabited a physically older avatar were subsequently more likely to agree to contribute more to their pensions.[28]

Time and again the research has shown that the As If principle does indeed operate in the virtual world and that seeing an avatar look and behave in a certain way affects how people think and behave in the real world. Bailenson has labeled the phenomenon the "Proteus effect," after the Greek god who can change his shape and identity at will. The effect opens up the As If principle to new worlds, limited only by our imaginations.

V. Turning Back Time

Born in the Bronx, psychologist Ellen Langer began her college career studying chemistry at New York University. Realizing that a life among the test tubes wasn't for her (she later noted, "I practiced Jewish chemistry—a little is good, more is better"), she enrolled in an introductory psychology course being taught by Philip Zimbardo and was hooked. Langer eventually became a professor at Harvard University, with much of her work attempting to unravel the mysteries of aging.

She has conducted many high-profile studies during her career. In one classic experiment, some of the residents in a nursing home were given a houseplant to look after, while the others were given an identical plant but told that the staff would take responsibility for it.[29] Six months later, the residents who had been robbed of even this small amount of control over their lives were significantly less happy, healthy, and active than the others. Even more distressing, 30 percent of the residents who had not looked after their plant had died, compared with just 15 percent of those who had been allowed to exercise control.

In a similar study, she examined the effect of encouraging elderly people to behave as if they are still mentally active.[30] In this experiment, the researchers visited one group of residents each week and asked them various questions, such as the names of the nurses and the nature of the activities that would be taking place in the nursing home on particular days. If the residents were uncertain about the answers, the experimenters encouraged them to find out by the next visit. The effects were dramatic. Compared with a control group of residents who were not presented with such challenges, those asked to remember simple questions developed superior short-term memories and appeared far more alert. The researchers revisited the homes two and a half years later and found that only 7 percent of the group had passed away, compared with nearly 30 percent of the control group.[31]

Perhaps Langer's most striking study involves getting people to use the As If principle to travel back in time.[32] In 1979 she recruited a group of men in their seventies and eighties for what was described as a week of reminiscence at a retreat outside Boston. Before the study started, Langer asked all the men to take part in a series of tests measuring their physical strength, posture, eyesight, and memory.

She then randomly split the men into two groups and told one group (the "time travelers") that the experiment was about the psychological impact of actually reliving the past and the other group that it was all about the effects of reminiscing. Langer had decided to try to turn back the hands of time by twenty years and so encouraged the participants to either relive or reminisce about their lives in 1959.

Next, she arranged to bus the group that would be reliving the past to a ten-acre country retreat. To get them in the mood for the study, Langer played 1959 radio broadcasts throughout the trip. She was interested in encouraging the participants to act as if they were twenty years younger throughout the experiment. When the time travelers arrived at the retreat, for instance, there was no one there to help them off the bus, and they had to carry their suitcases inside. In addition, the retreat had not been equipped with the types of rails and other movement aids they had at the nursing home.

Participants had been asked to supply photographs of themselves from 1959, and when they went to their bedrooms, each participant found copies of all of those photographs along with 1959 issues of *Life* magazine and the *Saturday Evening Post*.

After unpacking, everyone was assembled in the main room of the retreat. Surrounded by various objects from the era, including a black-and-white television and a vintage radio, Langer informed the participants that for the next few days, all of their conversations about the past had to be in the present tense and that no conversation must mention anything that happened after 1959.

Each day the participants took part in various carefully orchestrated activities and discussions. During one activity, for example, they were asked to write an autobiographical sketch in the present tense, ending in 1959. Other times they were taken to a makeshift movie theater to watch *Anatomy of a Murder* starring James Stewart, took part in discussions about the launch of the first American satellite, played games such as *The Price Is Right* using old prices, heard a speech by President Eisenhower, and sat around the radio listening as the horse Royal Orbit won the 1959 Preakness Stakes.

For the control group, life was very different. They heard present-day music on the bus, were asked to reminisce about 1959 in the past tense, were given current photographs of one another, and watched recent movies.

Within days, Langer could see the dramatic effect of behaving As If. The time-traveling participants were now walking faster and were more confident. Moreover, within a week, several of these participants had decided that they could now manage without their canes. Langer took various psychological and physiological measurements throughout the experiment and discovered that the time-traveling group showed improvements in dexterity, speed of movement, memory, blood pressure, eyesight, and hearing. Interestingly, more than 60 percent of those in the time-traveling group showed an improvement on intelligence tests compared with just 40 percent of those in the other group. Acting as if they were young men had knocked years off their bodies and minds.

To discover if Langer's original work could be repeated, the BBC recently reran her experiment. Six elderly British celebrities agreed to try to turn back the clock to their heyday during the 1970s. The BBC uncovered photographs of the celebrities' bedrooms during the 1970s and reconstructed the rooms in minute detail, complete with psychedelic wallpaper and swirling carpets. During the week, each celebrity was given the opportunity to relive an important moment in his or her life, with, for example, dancer Lionel Blair returning to the stage of the Palladium to choreograph a routine.

Within just a day or two, many of the celebrities showed an improvement in their memories, physical strength, energy, and mood. The eighty-eight-year-old actress Liz Smith had had three immobilizing strokes, but soon showed that she was able to get around without her cane. Prior to the experiment, cricket umpire Dickie Bird had become solitary, but within days he was the life and soul of the party. Tests assessing the participants' biological age suggested that two of the group had the brain of a person twenty years their junior, and overall the group showed significant improvements in memory and intelligence.

This is not the only set of experiments to show that it is possible to slow down the effects of aging by acting like a younger person. In another experiment, Langer asked people to role-play air force pilots and examined the impact on their vision.[33] Nineteen air force cadets had their eyesight tested and were then split randomly into two groups. An instructor invited each of the cadets in one of the groups to go into a flight simulator and attempt to fly a plane. Those in the other group were also asked to climb into the pilot's seat but were told the simulator was broken. All of the participants were then asked to read the letters on the sides of airplanes they saw through the cockpit window. Those acting as if they were fighter pilots improved their vision by 40 percent, while those who weren't behaving in this way showed no change at all.

Other work examined whether being around children really does keep you young. In one study, Langer examined the life expectancies of women who had children later in life compared with those who had them in their younger years.[34] You might think that running around very young children when you are in your forties is not such a great idea. You would be wrong. In fact, the women who had children later in life had significantly longer life expectancies than others. In a similar vein, Langer looked through the marriage registers searching for couples who differed in age by more than four years.

Langer hypothesized that the younger partners will be likely to behave as if they are older than they actually are and the older partners as if they are younger. This had a significant effect on their life expectancy, with the younger ones having significantly shorter life spans.

Then, not for the first time in this book, there is the power of dance. Researchers at the Albert Einstein College of Medicine in New York City followed a group of more than five hundred participants between 1980 and 2001.[35] When they joined the study, everyone was to indicate the degree to which they carried out various activities that stimulated their brains (reading, writing for pleasure, solving crosswords, playing board games, taking part in discussions, and playing a musical instrument) or bodies (playing tennis or golf, swimming, bicycling, dancing, walking, climbing, doing housework). When five hundred of the participants were more than seventy-five years old, the researchers monitored the degree to which they had developed dementia. Those who were readers showed a 35 percent reduced risk of dementia, while doing crossword puzzles at least four times a week showed a 47 percent drop. Interestingly, almost none of the physical activities, such as bicycling and swimming had any impact at all. The exception to the rule was dancing, with the regular dancers in the group exhibiting a massive 76 percent drop. By consistently dancing the night away, these people were acting as if they were young, and this helped slow the effects of aging over time.

As the playwright George Bernard Shaw once wisely remarked, "We don't stop playing because we grow old; we grow old because we stop playing."

How to Slow Aging

Here are five tips, based on the research of Ellen Langer, to help slow the effects of aging:

Maintain a sense of control in your life. Do not associate old age with helplessness and reliance on others. Instead, try to exert control over as many aspects of your life as possible. Langer's work suggests that even the smallest amounts of control make a big difference. Shop for yourself, look after a houseplant, tend to your garden, get a pet, take charge of your finances, and get out and about under your own steam.

Keep mentally active. There is a great deal of debate about whether so-called brain training affects your mental well-being. However, acting as if you are interested in the world around you is beneficial. Keep up with world news, find out what is happening in your area, start a blog, set goals, remain curious, maintain hobbies and interests, and stay in touch with friends and family.

Stay young at heart. Langer's work shows that those who spend time with children and younger people stay young. Make room in your life for grandchildren and young friends and neighbors.

Be actively active. Try to move like a younger person. Keep as physically active as you can, continue to be involved in sports, put a spring in your step, and remember that by far the most psychological benefit is derived from dancing.

Make an effort. The way that you look influences how you feel. In one study, Langer measured the blood pressure of women before and after they had their hair colored.[36] Women who thought that they looked younger after the appointment showed a significant reduction in blood pressure. Make an effort to look and dress a few years younger than you actually are.

Conclusion

Where we hypnotize a woman, saw a brain in half, and discover why you really are of two minds about everything

"Everything we have seen indicates that the surgery has left these people with two separate minds. That is, two separate spheres of consciousness."

—Roger Sperry, neuroscientist

Our journey has revealed the remarkable truth about your body and brain. For thousands of years, people have assumed that the relationship between their brain and body is much like that of a rider to a horse. In the same way that a rider determines how a horse behaves, so our minds decide what our bodies do. Because of this, those wishing to change their lives have spent vast amounts of time and money trying to alter how they think. Encouraged by self-appointed gurus and life coaches, these people have attempted to visualize their perfect self, think like a millionaire, and adopt a positive mind-set. Unfortunately, this approach to change is difficult, time-consuming, and frequently ineffective.

More than a century ago Harvard philosopher William James turned the conventional view of the human psyche on its head. He suggested that our actions influence how we think and feel, and argued that it should be possible to easily alter our thoughts and emotions by changing our behavior. Eighty years after James proposed his odd-sounding theory, a small number of maverick researchers carried out the first experiments to discover whether he was right. Their positive findings inspired other scientists to conduct similar research, and over time this work has revealed that the As If principle can help explain people's emotions, levels of motivation, beliefs, and personalities.

The message from this remarkable body of research is clear: it is not just that the mind influences the body but rather that the body also influences the mind. This simple idea has given rise to a series of quick, easy, and effective techniques that can help people become happier, avoid anxiety and depression, fall in love and live happily ever after, beat procrastination, and even slow the effects of aging. None of these techniques is about trying to change the way you think. Instead, they all entail ripping up the rules about self-development, changing the way you behave, and are based on William James's century-old maxim, "If you want a quality, act as if you already have it."

This work has important consequences for anyone wishing to understand the mysteries of the mind. Since its inception around the turn of the last century, psychology has struggled to find a single concept that can be applied to the many different facets of the human psyche, including emotion, thought, and behavior. For example, research into the psychology of motivation has produced several theories that help explain what makes people get up and go, but these ideas do little to further our understanding of happiness. Similarly, other work has tackled what happens when people feel sad, but these theories have nothing to say to those interested in understanding the psychology of persuasion. The As If principle does not suffer from such limitations. From passion to phobias, confidence to creativity, and persistence to personality, the same simple notion provides a remarkable insight into a diverse range of phenomena and, in doing so, offers the very real possibility of being psychology's first unifying theory.

Although such theoretical considerations are fascinating, the As If principle also has important practical consequences. As we have seen throughout this book, this deceptively simple idea provides the basis for a diverse range of self-development techniques that are quick, easy, and highly effective. Clench your muscles, and you develop instant willpower, force your face into a smile and you feel happier, stand up straight and you become far more confident. The same principle has also inspired other exercises that pave the way for longer-lasting and large-scale change, including those that help people shape their personality, lose weight, and alter the beliefs of entire nations.

New experiments into the As If principle are now being regularly reported at scientific conferences and published in academic journals. Just over a hundred years after James came up with his controversial theory, the principle has come in from the cold and is accepted as part of mainstream psychology. Indeed, some researchers are convinced that the As If principle is not just a fascinating quirk of the human mind but is actually operating every moment of every human being's waking life. To support their case, they point to a strange body of work involving posthypnotic suggestion, brain surgery, and a picture of a chicken's claw.

Dr. Zomb and the People with Two Brains

When I was a child, I developed a strong interest in conjuring and have worked as a professional magician. When I was in my early twenties, I thought that it might be fun to learn how to become a hypnotist, and so I visited my local magic shop and bought a book entitled *The New Encyclopedia of Stage Hypnotism* by Ormond McGill.

McGill was an experienced hypnotist who had worked under the stage name Dr. Zomb, and his book claimed to present a step-by-step guide to placing people in a deep trance. Inspired by Zomb's authoritative tone, I read his encyclopedia from cover to cover, learned the various induction procedures, and decided to give it a try. Rather than risk making a fool of myself onstage, I thought it best to try out my newly found skills on Claire.

Claire and I had been sharing a house for about a year, and so she was used to being asked to choose cards and think of the first number that came into her head. I asked Claire whether she would like to be my very first hypnotism guinea pig, and she kindly agreed. About ten minutes later, Claire was lying on our comfy sofa, and I was sitting in a nearby chair. I recited one of Dr. Zomb's most powerful induction procedures, asking Claire to close her eyes and drift away from consciousness. After a few minutes, Claire was lying motionless on the sofa.

Everything seemed to be going well, so I asked Claire to imagine that the number six didn't exist and then to count from one to ten. A few moments later Claire started her count, and I was thrilled to hear her confidently come out with the numbers "four, five, seven, eight."

After working my way through several other standard hypnotic exercises, I decided to add one of Dr Zomb's postsession suggestions. I told Claire that when she awoke, she wouldn't remember anything about the sessions but would feel a strange urge to wander across the room and open a window.

A few moments later, I counted to ten and Claire suddenly opened her eyes. Slightly dazed, she asked how the session had gone. I said that all was well and explained about the mystery of the missing six and the other exercises. At the end of the discussion, Claire stood up, walked across the room, and opened the window. Bemused, I casually asked her why she had opened the window on what was quite a chilly day. Claire confidently explained that she felt hot and so needed some fresh air.

Many proponents of hypnosis would argue that it is as if there are two people inside Claire's head: one of them controlled her behavior (let's call him or her "The Boss") and the other observes these actions and tries to work out what is going on ("The Observer"). According to this theory, in everyday life The Boss is actually the one in charge, but people are aware of only The Observer. So, for example, The Boss might head to a restaurant, and The Observer would look at what was going on and conclude that he or she must be hungry. Or The Boss might look longingly at his or her spouse, and The Observer might conclude that they are in love.

However, when people are hypnotized The Observer takes some much-needed time off, and it is possible to talk directly to The Boss. When I hypnotized Claire, I told The Boss to open the window when the session had ended. When Claire came out of the trance, The Observer woke back up and all seemed well. Then, as instructed, The Boss opened the window, and The Observer saw what was going on and assumed that Claire must feel hot.

Psychologists have long debated the nature of hypnosis. Some have argued that people can indeed enter a strange trancelike state and that during a hypnotic session, it is possible to talk to other aspects of a person's psyche. In contrast, others have argued that hypnosis is just an elaborate form of role play.[1] It is a long and complex debate, and so my experience with Claire does not constitute strong proof for the existence of The Boss and The Observer. Fortunately, this is far from the only evidence that we all have two people inside our head.

During some brain operations, patients can be fully conscious and able to report their thoughts and feelings. In the late 1960s, brain surgeon José Delgado thought that it would be interesting to see how his patients felt when he stimulated different parts of their brain during an operation.[2] In several operations, Delgado applied a tiny electrical pulse to the part of the brain that controlled head turning. On cue, each of the patients slowly rotated his or her head to the right and then the left. However, when Delgado asked the patients why they were behaving in this way, most of them quickly justified their actions, explaining, for example, that they were looking for their slippers, had heard a noise, or just generally felt restless. Once again, it seemed that the patients had looked at their actions and created a plausible narrative to explain what was going on.

By far the most profound work into this notion was carried out in the early 1970s by an American neuropsychologist, Roger Sperry.

Sperry and his colleagues originally set out to devise a new treatment for epilepsy. They knew that the brain consists of two large hemispheres connected by a thick band of nerve fibers called the corpus callosum and that previous work had revealed that some epileptic episodes were the result of excessive electrical activity in one of these two hemispheres quickly spreading to the other. Sperry wondered if it might be possible to prevent this electrical storm by cutting the tissue that connected the two hemispheres. To find out, he operated on several patients with epilepsy, severing the entirety of their corpus callosum. This dramatic procedure proved a considerable success, with the majority of the patients going on to lead normal lives.

However, it was Sperry's additional experiments with these so-called split-brain patients that yielded a remarkable insight into who we are.

Each of the hemispheres controls muscles on the opposite side of the body. The right hemisphere controls muscles on the left side of your body, for instance, and the left hemisphere controls muscles on your right side. The same process also applies to the images entering your eyes, with images from the left-hand side of your visual field going to the right hemisphere and images from the right going to the left. Normally the corpus callosum ensures that this information quickly flows between the hemispheres, and so is available to both sides of the brain. Sperry realized that this sharing of information couldn't happen in his "split brain" patients, and thus that it should be possible to work with them to investigate the different abilities of the hemispheres.

The research team constructed a laboratory inside a camper-trailer, hitched the trailer up to a van, and toured around the country testing participants.

In one series of studies, each participant was asked to focus on a dot in the center of a screen. Sperry then flashed up images on either the right or left of the dot. These flashes were deliberately so fast that the patients did not have time to move their eyes, which in turn ensured that the image went to just the right or left hemisphere.

Language and self-awareness tend to be located in the left hemisphere, so when Sperry flashed up a picture of an object on the right of the screen, the patient could easily name the object. However, when the same picture was flashed on the left-hand side of the screen, it was fed to the right hemisphere, and the patient felt as if he or she had seen nothing. Nevertheless, the images influenced the patients' behavior. When, for example, the word *smile* was flashed up on the left of the screen, the patient would smile. When an image of a nude woman was flashed up to male patients, an appreciative grin would sometimes spread across their faces.

These patients had no insight into the real cause of their behavior, and so when asked to explain their actions, they would say that they were smiling because they found the experiment funny or said they looked "appreciative" because they found the experimenter attractive. Once again, it was as if The Observer was watching their actions and trying to explain what was going on.

In another study, the researchers showed the split-brain patients two pictures, presenting one picture to each hemisphere. The participants were then asked to look at an assortment of other pictures and chose an image that was most closely related to the original pictures. For example, in one instance, a patient had a picture of a chicken claw flashed to the left hemisphere and a snow scene to the right hemisphere. When shown the ensuing selection of other images, the patient correctly selected a shovel with his left hand and a chicken with his right hand. When the researchers asked the patient to account for his actions, the patient explained that the chicken claw was obviously associated with a chicken and that you needed a shovel to clean out a chicken shed. Once again, it was as if The Observer saw how the patient behaved, had no insight into why this was the case, and made up a story to account for his actions.

Sperry's work strongly suggests that one part of your brain decides whether to eat, sleep, laugh, or cry, and another observes your actions and then tells you a story in an attempt to explain what is going on. Seen in this way, the As If principle is not simply a fascinating quirk of the human mind, but instead underlies every thought and feeling that you have experienced throughout your entire life.

Over a century ago, William James proposed that behavior causes emotion. It was a simple idea that changed everything. A hundred years of research has revealed that James's theory applies to a remarkable range of psychological phenomena, from persuasion to procrastination, fear to phobias, and passion to personality. In addition, it provides a profound insight into the fundamental nature of the human mind, and has the power to change lives and shape societies. The time has come to jettison out-of-date ideas about the human psyche and adopt William James's radical theory. To use the power of the As If principle to help people improve their lives and to change the world.

Appendix

Ten Ways in Which Your Body Instantly Changes

Your Brain

The preceding chapters contain a series of quick and effective exercises that will help change the way you think and behave. Here is a one-stop description of ten of the most effective techniques set out in this book.

Motivation: Pull me—push you

Pushing an object away from you (and so behaving as if you don't like it) makes you dislike the object, whereas pulling it toward you (behaving as if you like it) makes you feel far more positive about it. Next time that you are confronted with some sugary snacks or chocolate cookies, simply push the plate away from you and feel the temptation fade.

Dieting: Using your nondominant hand

When you eat with your nondominant hand, you are acting as if you are carrying out an unusual behavior. Because of that, you place more attention on your action, do not simply consume food without thinking about it, and so eat less.

Willpower: Tensing up
Tensing your muscles boosts your willpower. Next time you feel the need to avoid a cigarette or piece of cake, make a fist, contract your bicep, press your thumb and first finger together, or grip a pen in your hand.

Persistence: Sit up straight and cross your arms
In several experiments researchers have presented volunteers with tricky problems and measured how long they persevered. Those who sit up straight and fold their arms persist for nearly twice as long as others. Make sure your computer monitor is slightly above your eye line, and when the going gets tough, cross your arms.

Confidence: Power posing
To increase your self-esteem and confidence, adopt a power pose. If you are sitting down, lean back, look up, and interlock your hands behind your head. If you are standing up, place your feet flat on the floor, push your shoulders back and your chest forward, and hold your hands in front of you.

Procrastination: Make a start
To overcome procrastination, act as if you are interested in what it is that you have to do. Spend just a few minutes carrying out the first part of whatever it is you are avoiding, and suddenly you will feel a strong need to complete the task.

Creativity: Move outside the box
If you want to come up with new ideas, act in a novel way. Spend some time walking around, but ensure that your path is curvy and unpredictable. If that doesn't get your creative juices flowing, try acting as if you are artistic by drawing, painting, or sculpting.

Persuasion: Getting the nod
Researchers have found that when people nod their heads up and down while they listen to a discussion (causing them to nod as if they agreed with the arguments), they are more likely to agree with the points being made. When you want to encourage others to agree with you, subtly nod your head as you chat with them. They will reciprocate the movement and find themselves strangely attracted to your way of thinking.

Negotiation: Warm tea and soft chairs

When people think that they are connected to others, they feel physically warm. And it is also true that when you warm someone up with a nice mug of tea or coffee, they become far friendlier. Similarly, hard furniture is associated with hard behavior. In one study, researchers had participants sit on either soft or hard chairs and then negotiate over the price of a used car. Those in the hard chairs offered less and were more inflexible.

Guilt: Wash away your sins

If you are feeling a tad guilty about something, try washing your hands or taking a shower. In several experiments, people who carried out an immoral act and then cleaned their hands with an antiseptic wipe felt significantly less guilty than others.

Acknowledgments

First and foremost, I thank the University of Hertfordshire for supporting my work over the years. I thank Clive Jefferies and Emma Greening for reading earlier drafts of the manuscript. This book would not have been possible without the guidance and expertise of my agent, Patrick Walsh, and editors Jon Butler and Millicent Bennett. Special thanks also to my wonderful colleague, collaborator, and partner, Caroline Watt. Finally, I would like to thank the many academics who dedicated their professional lives to creating and exploring the As If principle, including Daryl Bem, James D. Laird, Stanley Schachter, Arthur Aron, and, of course, that adorable genius, William James.

Notes

Chapter 1: How to Be Happy

1. Miller, G. A. (1993). *Psychology: The Science of Mental Life*. New York: Harper.
2. Much of the information in this section is based on *Hunt*, M. (1993). *The History of Psychology*. New York. Doubleday. Richardson, R. D. (2006). *William James: In the Maelstrom of American Modernism*. Boston: Houghton Mifflin.
3. Miller, G. A. (1962). *Psychology: The Science of Mental Life*. New York: Harper.
4. Cited in Myers, G. R. (1986). *William James: His Life and Thought*. New Haven, CT: Yale University Press.
5. James, W. (1892). *Psychology: Briefer Course*. New York: Holt.
6. William J. (1890). *The Principles of Psychology*, Vol. 1. New York: Holt.
7. Snyder, P. S., Kaufman, R., Harrison, J., & Maruff, P. (2010). Charles Darwin's emotional expression "experiment" and his contribution to modern neuropharmacology. *Journal of the History of the Neurosciences*, 19, 158–170.
8. James, W. (1899). *The Gospel of Relaxation*. New York: Scribner's.
9. Much of the information in this section is based on Laird, J. D. (2007). *Feelings: The Perception of Self*. New York: Oxford University Press.
10. Zajonc, R. B., Murphy, S. T., & Inglehart, M. (1989). Feeling and facial efference: Implications of the vascular theory of emotion. *Psychological Review*, 96(3), 395–416.
11. Larsen, R. J., Kasimatis, M., & Frey, K. (1992). Facilitating the furrowed brow: An unobtrusive test of the facial feedback hypothesis applied to unpleasant affect. *Cognition and Emotion*, 6, 321–338.

12. Strack, F., Martin, L. L., & Stepper, S. (1988). Inhibiting and facilitating conditions of the human smile: A nonobstrusive test of the facial feedback hypothesis. *Journal of Personality and Social Psychology*, 54, 768–777.

13. Levenson, R. W., Ekman, P., & Friesen, W. V. (1990). Voluntary facial action generates emotion-specific autonomic nervous system activity. *Psychophysiology*, 27(4), 363–384.

14. Levenson, R. W., Ekman, P., Heider, K., & Friesen, W. V. (1992). Emotion and autonomic nervous system activity in the Minangkabau of West Sumatra. *Journal of Personality and Social Psychology*, 62(6), 972–988.

15. Feinstein, J. S., Adolphs, R., Damasio, A., & Tranel, D. (2010). The human amygdala and the induction and experience of fear. *Current Biology*, 21, 34–38.

16. Lee, T. W., Josephs, O., Dolan, R. J., & Critchley, H. D. (2006). Imitating expressions: Emotion-specific neural substrates in facial mimicry. *Social Cognitive Affective Neuroscience*, 1, 122–335.

17. Hill, A., Rand, D., Nowak, M., & Christakis, N. (2010). Emotions as infectious diseases in a large social network: The SISa model. *Proceedings of the Royal Society B: Biological Sciences*, 277(1701), 3827–3835.

18. Snodgrass, S. E., Higgins, J. G., & Todisco, L. (1986). *The Effects of Walking Behavior on Mood*. Paper presented at the American Psychological Association convention.

19. Koch, S. C. (2011). Basic body rhythms and embodied intercorporality: From individual to interpersonal movement feedback. In W. Tschacher & C. Bergomi (Eds.), *The Implications of Embodiment: Cognition and Communication* (pp. 151–171). Exeter: Imprint Academic.

20. Velten, E. (1968). A laboratory task for induction of mood states. *Behavior Research and Therapy*, 6, 473–482.

21. Larsen, R. J., & Sinnett, L. M. (1991). Meta-analysis of experimental manipulations: Some factors affecting the Velten mood induction procedure. *Personality and Social Psychology*, 17, 323–334.

22. Hatfield, E., Hsee, C. K., Costello, J., Weisman, M. S., & Denney, C. (1995). The impact of vocal feedback on emotional experience and expression. *Journal of Social Behavior and Personality*, 10, 293–313.

23. Neuhoff, C. C., & Schaefer, C. (2002). Effects of laughing, smiling and howling on mood. *Psychological Reports*, 91, 1079–1080.

24. Kim, S., & Kim, J. (2007). Mood after various brief exercise and sport modes: Aerobics, hip-hop dancing, ice skating, and body conditioning. *Perceptual and Motor Skills*, 104, 1265–1270.

25. Lovatt, P. (2011). Personal communication. November 9, 2011.

26. Clift, S., Hancox, G., Morrison, I., Hess, B., Kreutz, G., & Stewart, D. (2010) Choral singing and psychological wellbeing. *Journal of Applied Arts and Health*, 1, 19–34.

27. Kreutz, G., Bongard, S., Rohrmann, S., Grebe, D., Bastian, H., & Hodapp V. Does singing provide health benefits? In *Proceedings of the 5th Triennial ESCOM Conference* (pp. 216–219).

Chapter 2: Attraction and Relationships

1. Riley, J. (1967, March 17). The saga of the Barefoot Bag on campus. *Life*, 72–73.

2. Cited in Berscheid, E., & Walster, E. (1978). *Interpersonal attraction*. Reading, MA: Addison-Wesley.

3. Clark, R. D., III, & Hatfield, E. (1989). Gender differences in receptivity to sexual offers. *Journal of Psychology and Human Sexuality*, 2, 39–55.

4. For a review of this work, see Sternberg, R. J., & Weiss, K. (2006). *The New Psychology of Love*. New Haven, CT: Yale University Press.

5. Byrne, D. (1971). *The Attraction Paradigm*. New York: Academic Press.

6. Hatfield, E. (1985). Passionate and companionate love. In R. J. Sternberg & M .L. Barnes (Eds.), *The Psychology of Love* (pp. 191–217). New Haven, CT: Yale University Press.

7. Traupmann, J., & Hatfield, E. (1981). Love and its effect on mental and physical health. In R. Fogel, E. Hatfield, S. Kiesler, & E. Shanas (Eds.). *Ageing: Stability and Change in the Family* (pp. 253–274). New York: Academic Press.

8. These questionnaires are based on Hatfield, E., & Sprecher, S. (1986). Measuring passionate love in intimate relationships. *Journal of Adolescence*, 6, 383–410. Sprecher, S., & Fehr, B. (2005). Compassionate love for close others and humanity. *Journal of Social and Personal Relationships*, 22, 629–652.

9. Beauman, F. (2011). *Shapely Ankle Preferr'd: A History of the Lonely Hearts Ad, 1695–2010*. London: Chatto & Windus.

10. Asendorpf, J. B., Penke, L., & Back, M. D. (2011). From dating to mating and relating: Predictors of initial and long-term outcomes of speed-dating in a community sample. *European Journal of Personality*, 25, 16–30.

11. Schachter, S., & Singer, J. E. (1962). Cognitive, social and physiological determinants of emotional states, *Psychological Review*, 69, 379–399.

12. This questionnaire is based on the Body Perception Questionnaire by Stephen Porges.

13. Anderson, C. A., Bushman, B. J., & Groom, R. W. (1997). Hot years and serious and deadly assault: Empirical tests of the heat hypothesis. *Journal of Personality and Social Psychology*, 73, 1213–1223.

14. Baron, R. A., & Bell, P. A. (1976). Aggression and heat: The influence of ambient temperature negative affect, and a cooling drink on physical aggression. *Journal of Personality and Social Psychology*, 33, 245–255.

15. White, G. L., Fishbein, S., & Rutstein, J. (1981). Passionate love and the misattribution of arousal. *Journal of Personality and Social Psychology*, 41, 56–62.

16. For a review of this work, see Berscheid, E., & Walster, E. (1978). *Interpersonal Attraction*. Reading, MA: Addison-Wesley.

17. Dutton, D. G., & Aron, A. P. (1974). Some evidence for heightened sexual attraction under conditions of high anxiety. *Journal of Personality and Social Psychology*, 30, 510–517.

18. Meston, C. M., & Frohlich, P. F. (2003) Love at first fright: Partner salience moderates roller coaster-induced excitation transfer. *Archives of Sexual Behavior*, 32, 537–544.

19. Driscoll, R. H., Davis, K. E., & Lipetz, M. E. (1972). Parental interference and romantic love: The Romeo and Juliet effect. *Journal of Personality and Social Psychology*, 24, 1–10.

20. Hatfield, E., & Berscheid, E. (1971). Adrenaline makes the heart grow fonder. *Psychology Today*, 6, 46–62.

21. Brookes, M. (2004). *Extreme Measures: The Dark Visions and Bright Ideas of Francis Galton*. London: Bloomsbury.

22. Galton, F. (1884). Measurement of character. *Fortnightly Review*, 36, 179–182.

23. For a review of this work, see Givens, D. (2005). *Love Signals*. New York: St. Martin's Press.

24. Gergen, K., Gergen, M., & Barton, W. H. (1973). Deviance in the dark. *Psychology Today*, 11, 129–130.

25. Wegner, D. M., Lane, J. D., & Dimitri, S. (1994). The allure of secret relationships. *Journal of Personality and Social Psychology*, 66, 287–300.

26. Fraley, B., & Aron, A. (2004). The effect of a shared humorous experience on closeness in initial encounters. *Personal Relationships*, 11, 61–78.

27. For a review of this work, see Laird, J. D. (2007). *Feelings: The Perception of Self.* New York: Oxford University Press.

28. Epstein, R. (2006). Giving psychology away: A personal journey. *Perspectives on Psychological Science*, 1(4), 389–400.

29. Epstein, R. (2010, January/February). How science can help you fall in love. *Scientific American Mind*, 26–33.

30. Epstein, R. (2002, May/June). Editor as guinea pig: Putting love to a real test [Editorial]. *Psychology Today*, 5.

31. Much of the information in the section is from Epstein, R. (2005). Answers to 50 questions from the media: Regarding Dr. Epstein's "Learning-to-Love" experiment. http://drrobertepstein.com/downloads/Q&A-Learning_to_Love-4-05-c_2005_Dr._Robert_Epstein.pdf

32. See Goodwin, J. (2010, April 10). Love put to test in psych course. *U-T San Diego.* www.signonsandiego.com/news/2009/apr/10/1n10psych232525-love-put-test-psych-course/

33. Epstein, R. (2010, January/February). How science can help you fall in love. *Scientific American Mind*, 26–33.

34. Strong, G., Fincham, F. , & Aron, A. (2009). When nothing bad happens but you're still unhappy: Boredom in romantic relationships. *Mind*, 8. Tsapelas, I., Aron, A., & Orbuch, T. (2009). Marital boredom now, predicts less satisfaction nine years later. *Psychological Science*, 20, 543–545.

35. Reissmann, C., Aron, A., & Bergen, M. (1993). Shared activities and marital satisfaction: Causal direction and self-expansion versusboredom. *Journal of Social and Personal Relationships*, 10, 243–254.

Chapter 3: Mental Health

1. Hohmann, G. W. (1966). Some effects of spinal cord lesions on experienced emotional feelings. *Psychophysiology*, 3, 143–156.

2. For a more complete discussion of these findings and the various attempts to replicate them, see Laird, J. D. (2007). *Feelings: The Perception of Self*. New York: Oxford University Press.

3. Davis, J. I., Senghas, A., Brandt, F., & Ochsner, K. N. (2010). The effects of Botox injections on emotional experience. *Emotion*, 10(3), 433–440.

4. Cited in Chaves, J., & Barber, T. X. (1973). Needles and knives: Behind the mystery of acupuncture and Chinese meridians. *Human Behavior*, 2, 19–24.

5. Lanzetta, J., Cartwright-Smith, J., & Kleck, R. (1976). Effects of nonverbal dissimulation on emotional experience and autonomic arousal. *Journal of Personality and Social Psychology*, 3, 354–370.

6. Bohns, V. K., & Wiltermuth, S. S. (2012). It hurts when I do this (or you do that): Posture and pain tolerance. *Journal of Experimental Social Psychology*, 48, 341–345.

7. U.S. Federal Bureau of Investigation. (2010). *Uniform crime reports*. Washington, DC: U.S. Government Printing Office.

8. Cited in Seligman, Martin E .P. (1993). *What You Can Change and What You Can't: The Complete Guide to Successful Self-Improvement*. New York: Knopf.

9. Much of the information in this section is taken from Alvarado, C. (2009). Nineteenth-century hysteria and hypnosis: A historical note on Blanche Wittmann. *Australian Journal of Clinical and Experimental Hypnosis, 37*, 21–36.

10. Simon, R. I. (1967). Great paths cross: Freud and James at Clark University, 1909. *American Journal of Psychiatry*, 124(6), 139–142.

11. Straus, M. A. (1974). Leveling, civility, and violence in the family. *Journal of Marriage and the Family*, 36, 13–30.

12. Ebbesen, E. B., Duncan, B., & Kone?ni, V. J. (1875). Effects of content of verbal aggression on future verbal aggression: A field experiment. *Journal of Experimental Social Psychology*, 11, 192–204.

13. Goldstein, J. H., & Arms, R. L. (1971). Effects of observing athletic contests on hostility. *Sociometry*, 34, 83–90.

14. Johnson, S. (2011, March 11). Revealed: Full scale of old firm football "madness." *Telegraph*. www.telegraph.co.uk/news/uknews/scotland/8376041/Revealed-full-scale-of-Old-Firm-football-madness.html

15. Whitaker, J. L., & Bushman, B. J. (2011). Remain calm. Be kind. Effects of relaxing video games on aggressive and prosocial behavior. *Social Psychological and Personality Science*, 3, 88-89.

16. Bremner, R. H., Koole, S. L., & Bushman, B. J. (2011). Pray for those who mistreat you: Effects of prayer on anger and aggression. *Personality and Social Psychology Bulletin*, 37, 830–837.

17. Bushman, B. J. (2002). Does venting anger feed or extinguish the flame? Catharsis, rumination, distraction, anger, and aggressive responding. *Personality and Social Psychology Bulletin*, 28, 724–731.

18. Much of the information in this section is taken from Hunt, M. (1993). *The Story of Psychology*. New York: Doubleday.

19. Watson, J. B. (1930). *Behaviorism*. Chicago: University of Chicago Press.

20. Gustafson, C. R., Kelly, D. J, Sweeney, M., & Garcia, J. (1976). Prey-lithium aversions: I. Coyotes and wolves. *Behavioral Biology*, 17, 61–72.

21. Cited in Seligman, Martin E. P. (1993). *What You Can Change and What You Can't: The Complete Guide to Successful Self-Improvement*. New York: Knopf.

22. See, for example, Clark, D. M. (1986). A cognitive approach to panic. *Behavior Research and Therapy*, 24, 461–470. Clark, D. M., Salkovskis, P. M., Hackmann, A., Wells, A., Ludgate, J., & Gelder, M. (1999). Brief cognitive therapy for panic disorder: A randomized controlled trial. *Journal of Consulting and Clinical Psychology*, 67, 583–589.

23. Clark, D. M., Salkovskis, P. M., Hackman, A., Middleton, H., Anastasiades, P., & Gelder, M. (1994). A comparison of cognitive therapy, applied relaxation and imipramine in the treatment of panic disorder. *British Journal of Psychiatry*, 164, 759–769.

24. Goldfried, M. R., Linehan, M. M., & Smith, J. L. (1978). Reduction of test anxiety through cognitive restructuring. *Journal of Consulting and Clinical Psychology* 46, 32–39. Goldfried, M. R., & Sobocinski, D. (1975). Effect of irrational beliefs on emotional arousal. *Journal of Consulting and Clinical Psychology*, 43, 504–510. Langer, T., Janis, I., & Wolfer, J. (1975). Reduction of psychological stress in surgical patients. *Journal of Experimental Social Psychology*, 11, 155–165.

25. Schnall, S., Haidt, J., Clore, G. L., & Jordan, H. (2008). Disgust as embodied moral judgment. *Personality and Social Psychology Bulletin*, 34(8), 1096–1109.

26. Zhong, C., & Liljenquist, K. (2006). Washing away your sins: Threatened morality and physical cleansing. *Science*, 313, 1451–1452.

27. Ben-Noun, L. (2003). What was the mental disease that afflicted King Saul? *Clinical Case Studies*, 2, 270–282.

28. Huisman, M. (2007). King Saul, work-related stress and depression. *Journal of Epidemiology and Community Health,* 61, 890.

29. Myers, D. (2001). *Exploring Psychology*. New York: Worth.

30. Cuijpers, P., van Straten, A., van Oppen, P., & Andersson, G. (2008). Are psychological and pharmacologic interventions equally effective in the treatment of adult depressive disorders? A meta-analysis of comparative studies. *Journal of Clinical Psychiatry*, 69, 1675–1685. Imel, Z. E., McKay, K. M., Malterer, M. B., & Wampold. B. E. (2008). A meta-analysis of psychotherapy and medication in unipolar depression and dysthymia. *Journal of Affective Disorders*, 110, 197–206.

31. Schnall, S., & Laird, J. D. (2003). Keep smiling: Enduring effects of facial expressions and postures on emotional experience and memory. *Cognition and Emotion*, 17, 787–797.

32. VanSwearingen, J. M., Cohn, J. F., & Bajaj-Luthra, A.(1999). Specific impairment of smiling increases the severity of depressive symptoms in patients with facial neuromuscular disorders. *Aesthetic Plastic Surgery*, 23, 416–423.

33. Koch, S. C., Morlinghaus, K., & Fuchs, T. (2007). The joy dance. Effects of a single dance intervention on patients with depression. *Arts in Psychotherapy*, 34, 340–349.

34. Lewinsohn, P. M. (1974). A behavioral approach to depression. In R. M. Friedman & M. M. Katz (eds.), *The Psychology of Depression: Contemporary Theory and Research*. New York: Wiley. Lewinsohn, P. M., Antonuccio, D. O., Breckenridge, J. S., & Teri, L. (1984). *The "Coping with Depression" Course*. Eugene, OR: Castalia.

35. For further information about behavioral activation, see Veale, D., & Willson, R. (2007). *Manage Your Mood: Using Behavioral Activation Techniques to Overcome Depression*. London: Robinson.

36. Dimidjian, S., Hollon, S. D., Dobson, K. S., Schmaling, K. B., Kohlenberg, R., Addis, M., et al. (2006). Randomized trial of behavioral activation, cognitive therapy, and antidepressant medication in the acute treatment of adults

with major depression. *Journal of Consulting and Clinical Psychology* 74(4), 658–670.

37. For a review of this work, see Spates, C. R., Pagoto, S., & Kalata, A. (2006). A qualitative and quantitative review of behavioral activation treatment of major depressive disorder. *Behavior Analyst Today*, 7(4), 508–518.

Chapter 4: Willpower

1. Kohn, A. (1993). *Punished by Rewards: The Trouble with Gold Stars, Incentive Plans, A's, Praise, and Other Bribes.* Boston: Houghton Mifflin.

2. Curry, S. J., Wagner, E. H., & Grothaus, L. C. (1991). Evaluation of intrinsic and extrinsic motivation interventions with a self-help smoking cessation program. *Journal of Consulting and Clinical Psychology*, 59, 318–324.

3. Geller,E. S., Rudd,J. R., Kalsher,M. J., Streff,F. M., & Lehman,G. R. (1987). Employer-based programs to motivate safety belt use (a review of short-term and long-term effects). *Journal of Safety Research*, 18, 1–17.

4. See, for example, McQuillan, J. (1997). The effects of incentives on reading. *Reading Research and Instruction*, 36, 111–125.

5. Cited in Kohn, A. (1993). *Punished by Rewards: The Trouble with Gold Stars, Incentive Plans, A's, Praise, and Other Bribes.* Boston: Houghton Mifflin.

6. Amabile, T. M. (1985). Motivation and creativity: effects of motivational orientation on creative writers. *Journal of Personality and Social Psychology*, 48, 393–399.

7. Deci, E. L. (1971). Effects of externally mediated rewards on intrinsic motivation. *Journal of Personality and Social Psychology*, 18, 105–115.

8. Lepper, M. R., Greene, D., & Nisbett, R. E. (1973). Undermining children's intrinsic interest with extrinsic reward: A test of the "overjustification" hypothesis. *Journal of Personality and Social Psychology*, 28, 129–137.

9. Janis, I. L., & Mann, L. (1965). Effectiveness of emotional role playing in modifying smoking habits and attitudes. *Journal of Experimental Research* in *Personality*, 1, 84–90.

10. Mann, L., & Janis, I. L. (1968). A follow-up study on the long-term effects of emotional role-playing. *Journal of Personality and Social Psychology*, 8, 339–342.

11. Pliner, P., Hart, H., Kohl, J., & Saari, D. (1974). Compliance without pressure: Some further data on the foot-in-the-door technique. *Journal of Experimental Social Psychology*, 10, 17–22.

12. Beaman, A .L., Cole, M. C., Preston, M., Klentz, B., & Mehrkens-Steblay, N. (1983). Fifteen years of foot-in-the-door research: A meta-analysis. *Personality and Social Psychology Bulletin*, 9, 181–196.

13. Meineri, S., & Guéguen N. (2008). An application of the foot-in-the-door strategy in the environmental field. *European Journal of Social Sciences*, 7, 71–74.

14. Guéguen, N., & Jacob, C. (2001). Fund-raising on the web: The effect of an electronic foot-in-the-door on donation. *CyberPsychology and Behavior*, 4, 705–709.

15. Guéguen, N., Pascual, A., Marchand, M., & Lourel, M. (2008). Foot-in-the-door technique using a courtship request: A field experiment. *Psychological Reports*, 103, 529–535.

16. Cialdini, R. B., Cacioppo, J.T ., Basset, R., & Miller, J. A. (1978). The low-ball procedure for producing compliance: Commitment then cost. *Journal of Personality and Social Psychology*, 36, 463–476.

17. Zeigarnik, B. V. (1927). Über das Behalten von erledigten und unerledigten Handlungen (The retention of completed and uncompleted activities). *Psychologische Forschung*, 9, 1–85.

18. Staub, E. (1989). *The Roots of Evil*. Cambridge: Cambridge University Press.

19. Purcell, A. H. (1981). The world's trashiest people: Will they clean up their act or throw away their future? *Futurist*, 2, 51–59.

20. Burn, S. M., & Oskamp, S. (1986). Increasing community recycling with persuasive communication and public commitment. *Journal of Applied Social Psychology*, 16, 29–41.

21. Martin, S. (1993). *What You Can Change and What You Can't: The Complete Guide to Successful Self-Improvement*. New York: Knopf.

22. Mann, T., Tomiyama, A. J., Westling, E., Lew, A., Samuels, B., & Chatman, J. (2007). Medicare's search for effective obesity treatments: Diets are not the answer. *American Psychologist*, 62, 220–333.

23. Förster, J. (2003). The influence of approach and avoidance motor actions on food intake. *European Journal of Social Psychology*, 33, 339–350.

24. Webber, L. S., Catellier, D .J., Lytle, L. A., Murray, D. M., Pratt, C. A., Young, D. R., et al. (2008). Promoting physical activity in middle-school girls: Trial of activity for adolescent girls. *American Journal of Preventative Medicine*, 34, 173–184.

25. Schachter, S. (1968). Obesity and eating: internal and external cues differentially affect the eating behavior of obese and normal subjects. *Science*, 161, 751–756.

26. Nisbett, R. (1968). Determinants of food intake in obesity. *Science*, 159, 1254–1255.

27. Goldman, R., Jaffa, M., & Schachter, S. (1968). Yom Kippur, Air France, dormitory food, and eating behavior of obese and normal persons. *Journal of Personality and Social Psychology*, 10, 117–123.

28. Wansink, B. (2004). Environmental factors that increase the food intake and consumption volume of unknowing consumers. *Annual Review of Nutrition*, 24, 455–479.

29 Herman, C. P., Olmsted, M. P., & Polivy, J. (1983). Obesity, externality, and susceptibility to social influence: An integrated analysis. *Journal of Personality and Social Psychology*, 45, 926–934.

30. Sentyrz, S. M., & Bushman, B. J. (1998). Mirror, mirror on the wall, who's the thinnest one of all? Effects of self-awareness on consumption of fatty, reduced-fat, and fat-free products. *Journal of Applied Psychology*, 83, 944–949.

31. Neal, D., Wood, W., Wu, M., & Kurlander, D. (2011). The pull of the past: When do habits persist despite conflict with motives? *Personality and Social Psychology Bulletin*, 37, 1428–1437.

32. Riskind, J. (1984). They stoop to conquer: Guiding and self-regulatory functions of physical posture after success and failure. *Journal of Personality and Social Psychology*, 47, 479–493.

33. Hyung-il, A., Teeters, A., Wang, A., Breazeal, C., & Picard, R. (2007, September). *Stoop to conquer: Posture and affect interact to influence computer users' persistence.* Paper presented at the Second International Conference on Affective Computing and Intelligent Interaction4, Lisbon, Portugal.

34. Fletcher, B. C., & Stead, B. (2000). *(Inner) FITness and the FIT Corporation Smart Strategies.* London: International Thomson Press.

35. Hung, I. W., & Labroo, A. A. (2011). From firm muscles to firm willpower: Understanding the role of embodied cognition in self-regulation. *Journal of Consumer Research, 37*, 1046–1058.

36. Friedman, R., & Elliot, A. J. (2008). The effect of arm crossing on persistence and performance. *European Journal of Social Psychology, 38*, 449–461.

37. Fletcher, B. C., Page, N., & Pine, K. J. (2007). A new behavioral intervention for tackling obesity: Do something different. *European Journal of Nutraceuticals and Functional Foods, 18*, 8–10. Fletcher, B. C., Hanson, J., Pine, K. J., & Page, N. (2011). FIT–Do something different: A new psychological intervention tool for facilitating weight loss. *Swiss Journal of Psychology, 70*, 25–34. Fletcher, B. C., & Page, N. (2008). FIT Science for weight loss: A controlled study of the benefits of enhancing behavioral flexibility. *European Journal of Nutraceuticals and Functional Foods, 19*, 20–23.

Chapter 5: Persuasion

1. Korea: Twenty-three Americas. (1953, October 5). *Time.*

2. Aronson, E. (2004). *The Social Animal* (9th ed.). New York: Worth Publishers.

3. Kassarjian, H. H., & Cohen, J. B. (1965). Cognitive dissonance and consumer behavior. *California Management Review, 8*, 55–64.

4. Jones, E. E., & Kohler, R. (1958). The effects of plausibility on the learning of controversial statements. *Journal of Abnormal and Social Psychology, 57*, 315–320.

5. Bickman, L. (1972). Environmental attitudes and actions. *Journal of Social Psychology, 87*, 323–324.

6. Batson, C. D., Thompson, E. R., & Chen, H. (2002). Moral hypocrisy: Addressing some alternatives. *Journal of Personality and Social Psychology, 83*, 330–339. Batson, C. D. Thompson, E. R., Seuferling, G., Whitney, H., & Strongman, J. (1999). Moral hypocrisy: Appearing moral to oneself without being so. *Journal of Personality and Social Psychology, 77*, 525–537.

7. For a review of this work, see Maio, G. R., & Haddock, G. (2010). *The Psychology of Attitudes and Attitude Change.* London: Sage.

8. Peterson. A. V. Jr., Kealey, K. A., Mann, S. L., Marek, P. M., & Sarason, I. G. (2000). Hutchinson Smoking Prevention Project: Long-term randomized trial in school-based tobacco use prevention–results on smoking. *Journal of the National Cancer Institute, 92*, 1979–1991.

9. Wakefield, M., Terry-McElrath, Y., Emery, S., Saffer, H., Chaloupka, F., Szczypka, G., et al. (2006). Effect of televised, tobacco company-funded smoking prevention advertising on youth smoking-related beliefs, intentions and behavior. *American Journal of Public Health*, 96, 2154–2160.

10. *Monitor 2010: Fresh fruit and vegetable production, trade, supply and consumption monitor in the EU-27 (covering 2004–2009)*. Report produced by Freshfel Europe. www.dailymail.co.uk/health/article-1264937/Millions-spent-5-day-mantra-eating-LESS-vegetables.html

11. Hornik, R., Jacobsohn, L., Orwin, R., Piesse, A., & Kalton, G. (2008). Effects of the National Youth Anti-Drug Media Campaign on youths. *American Journal of Public Health*, 98, 2229–2236.

12. Clark, K. B., & Clark, MK. (1940). Skin color as a factor in racial identification of Negro preschool children. *Journal of Social Psychology*, 11, 159–169. Clark, K. B., & Clark, M. K. (1950). Emotional factors in racial identification and preference in Negro children. *Journal of Negro Education*, 19, 341–350.

13. For an account of this work, see Kleinke, C. L. (1978). *Self-Perception: The Psychology of Personal Awareness*. San Francisco: Freeman.

14. Halberstam, D. (1976). *The Best and the Brightest*. New York: Random House.

15. Elms, A. C. (1966). Influence of fantasy ability on attitude change through role playing. *Journal of Personality and Social Psychology*, 4, 36–43.

16. Janis, I. L., & King, B. T. (1954). The influence of role-playing on opinion change. *Journal of Abnormal Social Psychology*, 49, 211–218.

17. Laird, J. D. (2007). *Feelings: The Perception of Self*. New York: Oxford University Press.

18. Schein, E. H. (1956). The Chinese indoctrination program for prisoners of war. *Psychiatry*, 19, 149–172.

19. Myers, D. G. (2010). *Social Psychology* (10th ed.). New York: McGraw-Hill.

20. Chandler, J., & Schwarz, N. (2009). How extending your middle finger affects your perception of others: Learned movements influence concept accessibility. *Journal of Experimental Social Psychology*, 45, 123–128.

21. Wells, G. L., & Petty, R. E. (1980). The effects of overt head movements on persuasion: Compatibility and incompatibility of responses. *Basic and Applied*

Social Psychology, 1, 219–230. Tom, G., Pettersen, P., Lau, T., Burton, T., & Cook, J. (1991). The role of overt head movement in the formation of affect. *Basic and Applied Social Psychology*, 12, 281–289.

22. For more information about this exercise, see Peters, W. (1971). *A Class Divided: Then and Now*. New Haven, CT: Yale University Press.

23. Bloom, S. G. (2004). *Blue-eyes, brown-eyes: The experiment that shocked the nation and turned a town against its most famous daughter*. www.uiowa .edu/~poroi/seminars/2004-5/bloom/poroi_paper.pdf

24. For Jones's firsthand account of the experiment, see Jones, R. (1972). *The third wave, 1967: An account—Ron Jones*. libcom.org. http://libcom.org/history/the-third-wave-1967-account-ron-jones

25. Chaiken, S., & Baldwin, M. W. (1981). Affective-cognitive consistency and the effect of salient behavioral information on the self-perception of attitudes. *Journal of Personality and Social Psychology*, 41, 1–12.

26. Brehm, J. (1956). Post-decision changes in desirability of alternatives. *Journal of Abnormal and Social Psychology*, 52, 384–389.

27. Knox, R. E., & Inkster, J. A. (1968). Postdecision dissonance at post time. *Journal of Personality and Social Psychology*, 8, 319–323.

28. Zhong, C. B., & Leonardelli, G. J. (2008). Cold and lonely: Does social exclusion literally feel cold? *Psychological Science*, 19, 838–842.

29. Williams, L. E., & Bargh, J. A. (2008). Experiencing physical warmth promotes interpersonal warmth. *Science*, 322, 606–607.

30 Brehm, J. W. (1960). Attitudinal consequences of commitment to unpleasant behavior. *Journal of Abnormal and Social Psychology*, 60, 370–383.

31. Glass, D. C. (1964). Changes in liking as a means of reducing cognitive discrepancies between self-esteem and aggression. *Journal of Personality*, 32, 531–549.

32. Hatfield, E., Cacioppo, J. T., & Rapson, R. L. (1994). *Emotional Contagion*. Cambridge: Cambridge University Press.

33. See, for example, Friedman, H. S., & Riggio, R. (1981). The effect of individual differences in nonverbal expressiveness on transmission of emotion. *Journal of Nonverbal Behavior*, 6, 96–104.

34. Andréasson, P., & Dimberg, U. (2008). Emotional empathy and facial feedback. *Journal of Nonverbal Behavior*, 32, 215–224. Dimberg, U., Andréasson,

P., & Thunberg, M. (2011). Emotional empathy and facial reactions to facial expressions. *Journal of Psychophysiology*, 25, 26–31.

35. This questionnaire is based on scales described in Friedman, H. S., Prince, L. M., Riggio, R. E., & DiMatteo, M. R. (1980). Understanding and assessing nonverbal expressiveness: The Affective Communication Test. *Journal of Personality and Social Psychology*, 39, 333–351. Mehrabian, A., & Epstein, N. (1972). A measure of emotional empathy. *Journal of Personality*, 40, 525–543.

36. Sherif, M., Harvey, O. J., White, B. J., Hood, W. R., & Sherif, C. W. (1961). *The Robbers Grave Experiment: Intergroup Cooperation and Conflict*. Norman: University of Oklahoma.

37. Wiltermuth, S. S., & Heath, C. (2009). Synchrony and cooperation. *Psychological Science*, 20, 1–5.

Chapter 6: Creating a New You

1. For an excellent overview of this work, see Matthews, G. & Deary, I. J. (1998). *Personality Traits*. Cambridge: Cambridge University Press.

2. For a description of this study, see Ross, L., & Nisbett, R. E. (1991). *The Person and the Situation: Perspectives of Social Psychology*. New York: McGraw-Hill.

3. Hartshorne, H., & May, M. A. (1928). *Studies in the Nature of Character* (Vol. 2). New York: Macmillan.

4. Comer, R., & Laird, J.D. (1975). Choosing to suffer as a consequence of expecting to suffer: Why do people do it? *Journal of Personality and Social Psychology*, 32, 92–101.

5. Foxman, J., & Radtke, R. C. (1970). Negative expectancy and the choice of an aversive task. *Journal of Personality and Social Psychology*, 15, 253–257.

6. Kellerman, J., & Laird, J. D. (1982). The effect of appearance on self-perception. *Journal of Personality*, 50, 296–315.

7. Carney, D., Cuddy, A.J.C., & Yap, A. (2010). Power posing: Brief nonverbal displays affect neuroendocrine levels and risk tolerance. *Psychological Science*, 21, 1363–1368.

8. Schubert, T. W. (2004). The power in your hand: Gender differences in bodily feedback from making a fist. *Personality and Social Psychology Bulletin*, 30, 757–769.

9. Briñol, P., & Petty, R. E (2003). Overt head movements and persuasion: A self-validation analysis. *Journal of Personality and Social Psychology*, 84, 1123–1139.

10. Griffin, J. H. (196½010). *Black like Me*. San Antonio: WingsPress. For an account of Griffin's life, see Bonazzi, R. (1997). *Man in the Mirror: John Howard Griffin and the Story of Black like Me*. New York: Orbis Books.

11. Guéguen, N. (2009). Man's uniform and receptivity of women to courtshiprequest: Three field experiments with a firefighter's uniform. *European Journal of Social Sciences*, 12, 235–240.

12. Townsend, J. M., & Levy, G .D. (1990). Effects of potential partners' physical attractiveness and socioeconomic status on sexuality and partner selection. *Archives of Sexual Behavior*, 19, 149–164.

13. Green, W. P., & Giles, H. (1973). Reactions to a stranger as a function of dress style: The tie. *Perceptual and Motor Skills*, 37, 676.

14. Frank, M. G., & Gilovich, T. (1988). The dark side of self and social perception: Black uniforms and aggression in professional sports. *Journal of Personality and Social Psychology*, 54, 74–78.

15. Johnson, R. D., & Downing, L. L. (1979). Deindividuation and valence of cues: Effects of prosocial and antisocial behavior. *Journal of Personality and Social Psychology*, 37, 1532–1538.

16. Tenzel, J. H., & Cizanckas, V. (1973). The uniform experiment. *Journal of Police Science and Administration*, 1, 424. Tenzel, J. H., Storms, L., & Sweetwood, H. (1976). Symbols and behavior: an experiment in altering the police role. *Journal of Police Science and Administration*, 1, 21–27.

17. Leung, A. K., Kim, S., Polman, E., Ong, L., Qiu, L., et al. (2012). Embodied metaphors and creative "acts." *Psychological Science*.

18. For further information about this work, see www.prisonexp.org/links.htm. Zimbardo, P. G., Maslach, C., & Haney, C. (2000). Reflections on the Stanford Prison Experiment: Genesis, transformations, consequences. In T. Blass (Ed.), *Obedience to Authority: Current Perspectives on the Milgram Paradigm* (pp. 193–237). Mahwah, NJ: Erlbaum.

19. Roberts, B. W. (1997). Plaster or plasticity: Are adult work experiences associated with personality change in women? *Journal of Personality*, 65, 205–231.

20. Kohn, M. L., & Schooler, C. (1978). The reciprocal effects of the substantive complexity of work and intellectual flexibility: A longitudinal assessment. *American Journal of Sociology*, 84, 24–52.

21. For more information about these approaches, see Winter, D. A. (1987). Personal construct psychotherapy as a radical alternative to social skills training (pp. 107–123). In R. A. Neimeyer & G. J. Neimeyer (Eds.), *Personal Construct Therapy Casebook*. New York: Springer. Beail, N., & Parker, C. (1991). Group fixed role therapy: A clinical application. *International Journal of Personal Construct Psychology*, 4, 85–96. Lira, F. T., Way, W. R., McCullough, J. O., & Etkin, W. (1975). Relative effects of modeling and role playing in the treatment of avoidance behaviors. *Journal of Consulting and Clinical Psychology*, 43, 608–618.

22. Clore, G. L., & Jeffrey, K. M. (1972). Emotional role playing, attitude change, and attraction toward a disabled person. *Journal of Personality and Social Psychology*, 23, 105–111.

23. The exercises described here are designed to provide general insight into the sorts of techniques that psychologists use. If you believe that you have a serious problem in your life, consult a professional.

24. This list is based on work described in Peterson, C., & Seligman, M.E.P. (2004). *Character Strengths and Virtues: A Handbook and Classification*. Washington, D.C.: APA Press and Oxford University Press.

25. If you have decided to make this change, please feel free to get in touch with me.

26. Yee, N., Bailenson, J. N., & Ducheneaut, N. (2009). The Proteus effect: Implications of transformed digital self-representation on online and offline behavior. *Communication Research*, 36, 285–312.

27. Fox, J. A., & Bailenson, J. N. (2009). Virtual self-modeling: The effects of vicarious reinforcement and identification on exercise behaviors. *Media Psychology*, 12, 1–25.

28. Hershfield, H. E., Goldstein, D. G., Sharpe, W. F., Fox, J., Yeykelis, L., et al. (2011). Increasing saving behavior through age-progressed renderings of the future self. *Journal of Marketing Research*, 48, S23–S37.

29. Rodin. J., & Langer, J. E. (1997). Long-term effects of a control-relevant Intervention with the institutionalized aged. *Journal of Personality and Social Psychology*, 35(12), 897–902.

30. Langer, E. J., Rodin, J., Beck, P., Weinman, C., & Spitzer, L. (1979). Environmental determinants of memory improvement in late adulthood. *Journal of Personality and Social Psychology*, 37, 2003–2013.

31. Langer, E. J., Beck, P., Janoff-Bulman, R., & Timko, C. (1984). The relationship between cognitive deprivation and longevity in senile and non-senile elderly populations. *Academic Psychology Bulletin*, 6, 211–226.

32. For further information about this project, see Langer, E. (2009). *Counter Clockwise: Mindful Health and the Power of Possibility*. New York: Ballantine Books. Langer, E. (1989). *Mindfulness*. Reading, MA: Addison-Wesley. Hsu, L. M., Chung, J., & Langer, E. J. (2010). The influence of age-related cues on health and longevity. *Perspectives on Psychological Science*, 5, 632–648.

33. Langer, E., Djikic, M., Pirson, M., Madenci, A. & Donohue, R. (2010), Believing is seeing: Using mindlessness (mindfully) to improve visual acuity. *Psychological Science*, 21, 661–666.

34. Hsu, L. M., Chung, J., & Langer, E. J. (2010). The influence of age-related cues on health and longevity. *Perspectives on Psychological Science*, 5, 632–648.

35. Verghese, J., Lipton, R. B., Katz, M., Hall, C. B., Kuslansky, G., Derby, C. A., et al. (2003). Leisure activities and the risk of dementia in the elderly. *New England Journal of Medicine*, 348, 2508–2516.

36. Hsu, L. M., Chung, J., & Langer, E. J. (2010). The influence of age-related cues on health and longevity. *Perspectives on Psychological Science*, 5, 632–648.

Conclusion

1. For an accessible account of this debate, see Waterfield, R. (2002). *Hidden Depths: The Story of Hypnosis*. London: Pan Macmillan.

2. Delgado, J.M.R. (1969). *Physical Control of the Mind: Towards a Psychocivilized Society*. New York: Harper.

Index

Abu Ghraib prison, abuses at, 170, 176
accelerometer, 133
adrenaline, 48, 51
Aesop, 170
African Americans, segregation and,
    156–57, 202–3
aging:
    reminiscing vs. recreating the past
        in, 228–30
    slowing the effects of, xi, 227–32
agreeableness, 191
Albert B. (infant phobia subject),
    93–94
Albert Einstein College of Medicine,
    231
alcoholism, 41, 148, 150
Allport, Gordon, 190
Amabile, Teresa, 118–19
amygdala, and fear, 19–20
"Analysis of a Phobia in a Five-Year-
    Old Boy" (Freud), 92
Anatomy of a Murder, 229
Andréasson, Per, 178

anger, aggression, 19, 46, 50, 53, 55,
    79, 83–88
    adverse effects of, 83
    black clothing and, 204–5
    escalation of, 86–87
    experimental production of,
        180–82, 212–15
    Freud vs. James on, 85–86
    in relationships, 86
    and watching sports, 87–88
    in workplace, 86–87
anger management courses, 89
anxiety, 46, 75, 79, 140, 214, 215
    continuum of, 95
    misinterpreted as passion, 56–57
anxious extrovert (choleric personality),
    190
anxious introvert (melancholy
    personality), 190
appearance:
    in personality, 202–10
    in slowing effects of aging, 232
Aristotle, 33, 35

Arkansas State University, 206

Aron, Arthur, 54, 60–61

Aronson, Elliot, 184–86

As If principle:

in achieving and maintaining
happiness, xi, 1–31, 45

in attraction and relationships, xi,
33–71

in controlling anger, 83–88

in creating and maintaining love,
59–71

defined, 11

effect on body of, 18

in explaining depression, 106–14

in increasing confidence, xi,
194–201

in increasing willpower, xi, 115–43

James's theory as basis of, xi–xii,
8–12, 17, 50, 57, 75, 79,
157–58, 235–37, 241–42

in pain management, 80–82

in persuasion, 145–86

physiology and, 46–57

in producing passion, 53–55

in promoting and maintaining
mental health, xi, 73–114

rewards vs., 117–23

role playing in, 123

in self-improvement, 187–232

in slowing effects of aging, xi,
227–32

as unifying theory of psychology,
236

as universal, 18

in virtual world, 224–26

in weight control, xi, 131–43

Aspects of the Novel (Forster), 145

Association for Psychological Science,
6–7

astrology, 189

attraction and relationships:

As If principle in, xi, 33–71

effect of clothing on, 203–4

laboratory research into, 58–71

misattribution of bodily sensations
in, 46–57

nature of love in, 35–46

attribution, 103–6

Austin, Tex., 184–85

avatars, 224–26

avoidance, 108

Baez, Joan, 73

Bailenson, Jeremy, 224–26

Barnard College, 79

Barrymore, John, 35

Batson, Daniel, 151–52

BBC, 229

Beatty, Warren, 15, 63

behavior:

and bonding, 179–86

"Boss" and "Observer" in, 238–39

as cause of emotions, 8–12, 11, 17,
22–31, 50, 192, 193, 235–36,
241–42

emotions as cause of, 7–8, 8, 11,
11, 50

instinctive, 11

justification of, 170–78

memory and, 105–6

observation and measurement of, 91–92, 238–39, 241

observation of one's own, 10–11

saying vs. doing in, 150–54

unethical, 99; *see also* ethics, morality

behavioral activation, 106–14, 108–10, 111–13

behaviorists, 91–95

belief:

and brain, 156

changing of, 182–84

and justification of behavior, 170–78

persuasion and, 148–50, 159–62, 162–64

saying and, 150–52, 160–62

Bem, Daryl, 157–58

Ben-Gurion University of the Negev, 101

Benning, Annette, 63

Ben-Noun, Liubov, 101

Berlin, Irving, 38

Bible, 101

Bickman, Leonard, 150–51

Bird, Dickie, 230

Black Bag experiment, 36–40

black bile, 190

black clothing, 204–7

Black Like Me (Griffin), 202–3

Blair, Lionel, 230

blame, 104

blood, 190

blood pressure, 98

blood sugar, 134

bodily fluids, in determining personality, 190

bodily sensations:

emotions and, 50, 51

reinterpretation in panic attack of, 98–99

see also physiology

body mass index, 133

Bohns, Vanessa, 82

bonding, 37, 39, 184

camping project in, 179–84

experimental production of, 182–84

persuasion and, 176, 179–86

Book of Love, 65, 66–67

boredom, in relationships, 68

"Boss, The," 238–39

Botox (botulinum toxin), 78–79, 107

Boy Scouts, 128–29

brain:

beliefs and, 156

body and, 17–20, 235, 243–45

in depression, 102

hemispheres of, 239–40

in neural network, 75

in passionate love, 41

personality and, 191–92

in physiology of emotion, 50

rejuvenation of, 230, 231, 232

in relationship of behavior to emotion, 11

brain (*cont.*)
    split-, 240–41
    "two people in," 238–39, 241
    and unconscious, 83–84
brain surgery, 239–40
brainwashing, 147, 161
Brandeis University, 118
Brando, Marlon, 15
brass ball reflexive response test, 3–4,
    5, 6
breathing acceleration, 46, 47–48, 76
Brehm, Jack, 174
Bretagne, Université de, 203
Briñol, Pablo, 201
Brouillet, André, 83
Brown, Peter, 80, 82
Browning, Elizabeth Barrett, 35, 38
Bugsy, 62–63
Burn, Shawn, 128
Burton, Richard, 62
Bushman, Brad, 88–89
Byzantines, 43

California, University of, 17, 177
    at San Diego, 64, 86
California Polytechnic State University,
    128
calmness:
    exercise for, 89, 90
    power of, 88–90
cancer, smoking and, 149–50
Canterbury Christ Church University,
    31
Capilano River bridge experiment, 54

Carlos (Mexican American student),
    186
Carney, Dana, 197
Carson, Johnny, 166
Carthew, Anthony, 35
Castillo, Gabriella, 63–64
catastrophizing, 104
catharsis, 85–87
Cattell, Raymond, 190–91
celebrities:
    aging study on, 229–30
    love between, 62–63
Celtic (soccer team), 88
Cervantes, Miguel de, 30
change:
    through As If principle, xi–xii
    of belief, 182–84
    exercises for, xi–xii
    for the worse, 127
    see also specific exercises
Change4Life campaign, 129–30
Charcot, Jean-Martin, 83–84
charisma, charismatic communication,
    176–78
charity contributions, foot-in-the-door
    technique and, 124–25
Charles, Prince of Wales, 35
Charlotte (conspiracy theorist),
    148–49
children:
    rejuvenating effect of, 230, 232
    in research, 153, 162, 165–68, 174,
    180–83
China, pain management in, 80, 82

choleric (anxious extrovert) personality, 190

Cialdini, Robert, 126

civil rights movement, 156–57, 159

Claire (hypnotism subject), 237–39

Claremont, Calif., recycling in, 128

Clark, David, 98

Clark, Kenneth and Mamie, 156

Clark, Russell, 38–39

Clark University, 85, 106

classical conditioning, 93, 94

Cleopatra, 62

climate change, 150

closure, 65

clothing:
  black, 204–7
  effect on oneself of, 204–7
  effect on others of, 203–4

coffee breaks, 95

cognitive therapy (CT), 104–5, 113

Collection for the Improvement of Husbandry and Trade, The, 43–44

Colorado, University of, 56, 173, 197

Columbia University, 47, 135

communism, 147–48, 159, 160–61

compassionate love, 41, 43

competition:
  in education, 185
  hatred and, 180–82

computer games, anger and relaxation in, 88–89

computers:
  posture for perseverance at, 138
  virtual world of, 224–26

confidence:
  As If principle in developing, xi, 194–201
  language and, 23–24

Congress, U.S., in antidrug campaign, 154

conscientiousness, 191

consciousness, building blocks of, 4

consent, manufacturing of, 154–55, 168–69

conspiracy theories, 148–49

Cornell University, 204

corpus callosum, 239–40

cortisol, 199

creativity:
  behavior and, 207–10
  reward system and, 118–22

Cronkite, Walter, 37

crying, 53

dancing:
  and depression, 107
  happiness created through, 29–31
  rejuvenating effect of, 231, 232

darkness, attraction and, 59

Dartmouth College, 81

Darwin, Charles, 10, 92

David (biblical figure), 101

Davis, Joshua Ian, 79

death, 227, 228

Deci, Edward, 120–21

deep breathing, 90

Defense Department, U.S., 170

Delgado, José, 239

dementia, 231

De Niro, Robert, 15

depression, 75, 101–14, 214

    and anger, 83

    behavioral activation in, 108–10,
        111–13

    behavioral approaches in
        overcoming, 106–14

    brain in, 102

    conventional treatments for, 104–5

    frequency of, 102

    reversing downward cycle of, 108

    self-blame in, 104

    symptoms of, 102

desegregation, *see* segregation

desensitization, 95–97

desire, 50

"Deviance in the Dark" (Gergen), 59

Deyo, Yaacov, 44

Diana, Princess of Wales, 35

Dice Man, The (Rhinehart), 70–71

diet:

    healthy, 129–30, 142, 148, 150, 154

    low-calorie, 131

    see also eating; obesity; weight
        control

Dimidjian, Sona, 113

disabilities, 217

disgust, 99–100

distraction, in pain management, 82

dogs, in conditioning experiments,
    93, 94

"do something different" (DSD)
    techniques, 141–42

Douglass, Frederick, 115

dream analysis, 85

Driscoll, Richard, 56

drugs:

    antidepressant, 103, 105, 113

    campaigns against, 154

    use of, 41, 148, 160

Dr. Zomb (Ormond McGill), 237–38

Duchenne, Guillaume-Benjamin-
    Amand, 10

Duke University, 174

Dutton, Donald, 54

eating:

    As If principle and, 131–37, 139,
        140–41

    internal vs. external signals for,
        133–37

    of vegetables, 174

Ebbesen, Ebbe, 86–87

Eisenhower, Dwight D., 229

Ekman, Paul, 17–18

electrical shocks:

    in depression treatment, 102

    in research, 10, 53, 80–81, 175,
        206, 239

Eliot, Charles, 5

Elliot, Andrew, 141

Elliott, Jane, 165–66

emotions:

    behavior as cause of, 8–12, 11, 17,
        22–31, 50, 192, 193, 235–36,
        241–42

    as blunted in paraplegics, 77–78

as cause of behavior, 7–8, 8, 11, 11, 50
as contagious, 20–21, 177–78
Darwin's experiment on, 10
misattribution of, 46–57
paralysis and, 75–79
physiological effects of, 18–19, 46–57
as recognized in others, 8–10, 9
Wundt's theory of, 13
see also specific emotions
empathy, 176–78
for disabled, 217
endorphins, 30
environmental consciousness, 154–55, 168–69
epilepsy, brain in, 239–40
Epstein, Robert, 61–64
love project and proposed book of, 63–64
ethics, morality:
and guilt trips, 99–100
as inconsistent in children, 192
saying vs. doing in, 151–52, 153
in Zimbardo prisoner experiment, 214
excitement, 46
as cure for boring marriages, 69–71
exercise, 30
in aging, 231, 232
for health, 129–30
for weight control, 132–33, 142
Expression of the Emotions in Man and Animals, The (Darwin), 10

externals, 134–37
extroversion, 190, 191–92, 193
eye color experiment, 162, 165–66
Eysenck, Hans, 191

facial anatomy, 10
facial expressions:
in charismatic communication, 177–78
and depression, 107
emotions created by, 10, 17–18, 19, 50, 99–100, 107
manipulation of, 13–17, 19
in memory and behavior, 105
paralysis in, 78–79
factor analysis, 190
failure:
attribution of, 104
of marriages, 45
of persuasion campaigns, 148–50, 153–54
of rewards, 117–23, 142
of weight-control techniques, 131–33
Fairleigh Dickenson University, 27
farmers, 215
fasting, 135–36
Faust (Goethe), 1
fear, 19, 46
and fearlessness, 78
generalization of, 94
see also panic attacks; phobias
59 Seconds (Wiseman), xi–xii, 54–55, 89

fight-or-flight response, 48

financial rewards, 119–22

finger and thumb persuasion exercise, 162–63

Finzi, Eric, 107

Fish-Bird story, 55

fist, forming of, 199

fixed-role therapy, 216–17

Fletcher, Ben, 139–41, 141–42

flexibility, 139–40
    habit vs., 141–42

Florida Atlantic University, 22

Florida State University, 38

folkways, 156

football, American, 87–88, 205

foot-in-the-door technique, 124–29, 129–30

footsie, 60

Formation of Vegetable Mould, The (Darwin), 10

Forster, E. M., 145

fox and grapes fable, 171

Fraley, Barbara, 60–61

Frank, Mark, 204–5

free association, 85

Freeman, Walter, 102–3

Freud, Sigmund, 38–39, 189, 215
    on anger, 85–86, 89
    behaviorists vs., 92–93
    James vs., 85–86
    sexuality and, 85, 92, 94
    and unconscious mind, 84–85
    Watson vs., 91, 94
    see also psychoanalysis

Friedman, Howard, 177–78

Friedman, Ron, 141

friendship, friendliness, 39
    jigsaw method in creating, 185–86
    temperature and, 172–73

Froelich, Penny, 54

frontal lobotomy, 102–3

frowning, 13–14, 107, 177

frustration, misinterpreted as passion, 55–56

fun, value of, 22–31

furnishings, persuasion and, 164

fusion, 104

Galton, Francis, 58, 189

Geller, E. Scott, 118

"Gender Differences in Receptivity to Sexual Offers" (Hatfield and Clark), 39

genetics, in personality, 191

Gergen, Kenneth, 59

Germany, 3, 162, 168

Glasgow, Scotland, sports and aggression in, 88

Glass, David, 174–76

glasses experiment, 197

goal setting, in overcoming depression, 110–13

Goethe, Johann Wolfgang von, 1

Goetzinger, Charles, 36–37, 39

Golden Fleece Awards, 38

Golding, William, 181

Goldman, Ronald, 135–36

Goldstein, Jeffrey, 87–88

Goliath (biblical figure), 101
government, persuasion campaigns of,
        148, 150, 153–54, 159–62, 174
Graf, Herbert, *see* Little Hans
Great Britain:
    celebrity aging study in, 229–30
    government campaigns in, 154, 174
    happiness project in, 20–21
Greco-Turkish War, atrocities of, 179,
        182
Greece, military junta in, 127
grief, 78, 102
Griffin, John Howard, 202–3
Guéguen, Nicolas, 125, 203–4
guilt trips, 99–100
gymnastics, 88

habits, 139–40
    altering of, 141–42
haiku poetry, 119
Hale-Bopp comet, 149
Hamilton, Edith, 56
Hancox, Grenville, 31
hands, dominant and nondominant,
        200–201
handshaking, 23
hand washing, 100
handwriting, 200–201
happiness, 79
    As If principle in finding, xi, 1–31,
        45
    bodily sensations in, 51
    dancing and, 29–31
    laughter and, 26–31

mass participation project on,
        20–21
movement and, 22–26
music and singing and, 30–31
in real-world situations, 20–21
self-improvement exercises for,
        15–16, 25–26
simple idea for change in, 3–12
smiling and, 10–11, 13–17, 21, 22,
        107, 137, 158, 192, 194–95
talking oneself into, 25–26
testing a theory for, 13–21
value of fun in, 22–31
words and language and, 23–25,
        25–26
Harvard University, 5, 6, 44, 62, 179,
        186, 190, 227, 235
Hatfield, Elaine, 24, 38–39, 41
hatred, experimental production of,
        180–82
Hawaii, University of, 24
Hawthorne, Nathaniel, 1887
head nodding, 164
heart rate, 18, 46, 47–48, 50, 53, 57,
        76
    emotions produced by, 51–53
    love and, 54–55
Heath, Chip, 184
Heaven's Gate cult, 149
Heidelberg, University of, 107
height, and assertiveness, 224–26
"Heil Hitler," 162
hemispheres, brain, 239–40
Hertfordshire, University of, 30, 139

hip-hop dancing, 30–31

Hippocrates, 189–90

Hohmann, George, 77–78

hopelessness, 102

Hostage Barricade Database System, 57

hostages, 57

howling, 28–29

Hugo, Victor, 8

Huisman, Martijn, 101

Hung, Ris, 140

Hutchinson Smoking Prevention Project, 153

hypnosis, 82, 84, 85, 148, 160, 237–39

    author's experiment with, 237–38

identity, sense of:

    avatars and, 224–26

    behavior and, 193

    as changeable, 215–17, 217–23

    in competition, 182

    developing one's own, see self-development

    in eye-color experiment, 165–66

    loss of, 214

    in skin-color experiment, 202–3

imagery, 82

incentives, see rewards

Indonesia, 18

insanity, 83

instinct, 11

internals, 134–37

Internet:

    fantasy games on, 224–26

    in mass participation experimentation, 20–21

    online dating sites on, 44–45

    in popular culture, 39

introversion, 190, 191–92, 193

Iowa State University, 88, 215

Iraq, 170, 176

James, Henry, 6

James, William, vii, 6, 13, 16, 21, 28, 127

    birth and childhood of, 4–5

    as founding father of modern psychology, 6–8

    Freud vs., 85–86

    ill health of, 85

    and theory of emotion as cause of behavior, xi–xii, 8–12, 17, 50, 57, 79, 157–58, 235–37, 241–42

    wit of, 5

    Wundt vs., 5–6

jigsaw method, 185–86

job interviews, 189, 207

Johns Hopkins University, 91, 95

Johnson, Lyndon, 159

Johnson, Robert, 206

jokes, 26–27

Jolie, Angelina, 62

Jones, Ron, 166–68

Jung, Carl, 189

justification, 170–78

J. W. Goethe-Universität, 31

Kataria, Madan, 26–27

Kelly, George, 215–17, 217–18

Kennedy, John F., 103, 177

Kennedy, Rosemary, 103

Kim, Sungwoon, 30

King, Martin Luther, Jr., 165, 177, 185

Koch, Sabine, 107

Kohn, Alfie, 117

Korea, 30, 161

Korean War, persuasion of POWs in,
    147–48, 160–61

Korpi, Doug, 214

Kreutz, Gunter, 31

Ku Klux Klan outfit, 206

Kyungpook National University, 30

Laird, James, 13–14, 16, 106, 194–96

Langer, Ellen, 227–31, 231

Lanzetta, John, 81

laughter, happiness created through,
    26–31

laughter clubs, 26–28
    exercise for, 28–29

Leipzig, University of, 3

Lepper, Mark, 121–22

leucotome, 102–3

Leung, Angela, 208

Lewinsohn, Peter, 107–8

Li, Xiuping, 64

Lianne (pseudonym), 68

Lie to Me, 17

Life, 37, 228

liquid meals, 131

listening, singing vs., 31

littering, saying vs. doing in, 150–51,
    153

Little Hans (phobia subject), 92–95

loneliness, temperature and, 172–73

Lord of the Flies (Golding), 181

Lovatt, Peter (Dr. Dance), 30

love:
    attempts to define, 35
    barriers as enhancing, 55–56
    behaving as if in, 59–71
    chemistry of, 53–55
    determining the nature of, 35–46
    early research into, 36–40, 41
    evolution of, 41
    foot-in-the-door technique and,
        125–26
    inaccurate magical conceptions of,
        62
    laboratory research into, 58–71
    longevity of, 68–71
    passionate vs. compassionate,
        40–42
    questionnaire for, 42–43
    real-world experiments in, 61–64
    search for, 43–45
    as taboo topic, 37–38
    techniques for moving on, 64–65
    ubiquity of, 35–36
    unrequited, 55–57

love cake, 43

"love contract," 63

love games, 66–67

love letters, oldest, 35–36

love potions and spells, 43–44

low-ball technique, 126
low self-esteem, demeaning experience
    and, 194–96
lying, 17

McGill, Ormond (Dr. Zomb), 237
Mandela, Nelson, 177
manipulation, through persuasion,
    165–69
marriage:
    failures of, 45
    overcoming boredom in, 68–71
Martin, Steve, 54
Maryland, University of, 54
Maslach, Christina, 213–14
matchmaking, 44
Maxwell, Andrew, 148
maze experiments, 91–93
"Measurement of Character" (Galton),
    58
media:
    Black Bag experiment reported by, 37
    and Epstein's love project, 63
    eye-color experiment reported by,
        166
melancholy (anxious introvert)
    personality, 190
memory:
    in aging, 227–30
    and behavior, 105–6
    James's experiment on, 8
    and persuasion, 150
men:
    depression in, 102

in sex research, 39
Menlo Park, Calif., police uniforms in,
    206–7
mental health:
    As If principle in promoting and
        maintaining, xi, 73–114
    dealing with depression in, 101–14
    eliminating pain, anger, and anxiety
        in, 80–100
    paralysis and emotion in, 75–78
Meston, Cindy, 54
Mexican Americans, 186
Michigan, University of, 16, 157
Middle Ages, love in, 43
Milton, John, 8
mind control, 147–48, 160–61
mind reading, 104
mirror time technique, 215–16
moral hypocrisy, 151–52, 153
Mormons, love and, 36
motivation, see rewards; willpower
movement, emotions created through,
    22–26
movies, romance in, 62–63
Mr. & Mrs. Smith, 62
Mr. Angry, 52
Mr. Euphoric, 47, 51–52
muscles:
    in boosting willpower, 140–41
    brain's control of, 240–41
    in facial expression, 17–18
    hypothesis of memory as, 8
music, happiness created through,
    30–31

National Cancer Institute, U.S., 153

National Football League, 205

National Hockey League, 205

National Science Foundation, 38

Nazi ideology, 162

    experiment on, 166–68

neural network, 75–77

neurons, 103

neuroticism, 191

New Encyclopedia of Stage Hypnotism,
    The (McGill), 237

New Hampshire, University of, 86

New Orleans Superdome, 128

New York University, 227

"New You," creation of, *see* self-
    development

Nick (pseudonym), 68

Niffer Valley, Iraq, 35

Northwestern University, 172

nursing home residents, studies of,
    227–29

Obama, Barack, 177

obesity, 131–37

observation:

    brain in, 238–39, 241

    in love and attraction research, 58–59

    self-, 10–11

    Watson's techniques for, 91

"Observer, The," 238–39, 241

Ohio State University, 176–77

Oklahoma, 180

On the Origin of Species (Darwin), 10

openness, 191

Operation Match, 44

overconfidence, danger of, 172–74

Oxford, University of, 989

pain:

    inflicting of, 170, 174–78

    subjective nature of, 80–82

pain management, 75, 80–82

Palo Alto, Calif., 167, 212

panic attacks, 97–100

    reinterpretation of symptoms in,
    98–99

Paradise Lost (Milton), 8

paralysis:

    and emotion, 75–79

    in facial disorders, 107

    role playing of, 217

paraplegics, 78

paroxetine, 113

passionate love, 40–41, 42–43

    heart rate and, 54–55

patriotism, 162

Pavlov, Ivan, 93

permanence, 104

personal ads, 43–44

personality:

    creating one's own new, *see* self-
    development

    determining essence of, 189–93

    five dimensions of, 191

personality traits, 189–91

persuasion, 145–86

    from behavior to bonding in,
    179–86

persuasion (*cont.*)
  evidence vs. belief in, 148–50
  exercises for, 162–64
  manipulation of masses through,
    165–69
  manufacturing consent in, 154–55,
    168–69
  of oneself, *see* justification
  problems of, 147–55
  saying is believing in, 150–52,
    160–62
  in social interaction, 36–40
Philistines, 101
phlegm, 190
phlegmatic (relaxed introvert)
    personality, 190
phobias, 75, 92–95
  about horses, 92, 94
  techniques to overcome, 95–97
physiology, 18–19
  body talk questionnaire for, 49–50
  jukebox analogy for, 53
  misattribution in, 46–57
  of neural network, 75–77
  in obesity, 133–37
  red team-blue team analogy for,
    48, 51
  reinterpretation in panic attack of,
    97–99
  rejuvenation of, 229
pigeons, in reward theory, 117, 121
Pine, Karen, 141–42
Pitt, Brad, 62
Pittsburgh, University of, 107

Pliner, Patricia, 124
Plymouth, University of, 99
police, 206–7, 212
politics, persuasion and, 159–60
Pope, Alexander, 38
positive thinking, positive action vs.,
    xi–xii, 31, 235
posture:
  for confidence, 197, 198, 199
  inclination in, 58
  for motivation, 138
  powerful, 82
power poses, 197, 198, 199
prayer:
  and belief, 162
  calming power of, 89
preference, justification in, 170–74
Price is Right, The, 229
Principles of Psychology (James), 6, 13
prisoners:
  abuse of, 170, 176
  in Zimbardo's mock experiment,
    211–15
prisoners of war (POWs), persuasion
    of, 147–48, 160–61
procrastination, 126–27
progressive muscle relaxation, 90
Proteus effect, 226
Proxmire, William, 38
psychoanalysis, 84–85, 91, 92–93, 215
psychodrama, 217
psychology:
  As If principle as unifying theory
    of, 236

first laboratory experiment in, 3–4
first study of emotions in, 10
James as pioneer in, 5–7
love and relationships in, 37–38
Psychology Today, 62, 63
public opinion, shifts in, 158–59, 174
punching bags, 89
Punished by Rewards (Kohn), 117
punishment, 160, 161, 175–76
Pyramis and Thisbe myth, 56

racism:
    eye-color experiment in, 162,
        165–66
    skin-color experiment in, 202–3
    see also segregation, racial
Rangers (soccer team), 88
rats:
    in maze experiments, 91–93
    in phobia study, 93–94
"Rattlers" vs. "Eagles" experiment,
    179–83, 185
Rayner, Rosalie, 93–95
rebound effect, 56–57
recycling, foot-in-the-door technique
    in promoting, 128–29
reflexive response, 3–4
rejuvenation, xi, 227–32
relapses, 118
relaxation techniques, 82, 95
relaxed extrovert (sanguine personality),
    190
relaxed introvert (phlegmatic
    personality), 190

repression, 85
Requiem (Mozart), 31
restaurants, external temptations used
    by, 137–38
rewards:
    in competition, 181
    and creativity, 118–22
    failure of, 117–23, 142
    in mind control, 161
    as punishment, 119–22
    in workplace, 122
Rhinehart, Luke, 70
Riceville, Iowa, 165–66
Ring of the Nibelung, The (Wagner), 95
risk, 199
Riskind, John, 138
Rochester, University of, 13, 141
Rodney (conspiracy theorist), 148–49
role-playing, 123, 159–60
    through avatars, 224–26
    in developing new identity, 216–17
    in eyesight experiment, 230
    hypnosis as, 239
    in Zimbardo prisoner experiment,
        212–15
Romeo and Juliet (Shakespeare), 56

sadness, depression vs., 102
salesmanship, low-ball technique in,
    126
sanguine (relaxed extrovert) personality,
    190
Saturday Evening Post, 228
"Satyr" (Hugo), 8

Saul, King of Israel, as depressed, 101
Schachter, Stanley:
    obesity research of, 133–37, 139
    physiological research of, 45,
        46–53, 97–98
Schaefer, Charles, 27–28
Schnall, Simone, 99, 106
schools, segregated, 156–57
Schubert, Thomas, 199
seat belt use, 118, 174
Seattle, Wash., antismoking campaign
    in, 153
segregation:
    eye color experiment in, 165–66
    racial, 156–57, 158–59, 184–86
self-development, 187–232
    appearance in, 202–10
    confidence and, 194–201
    creating new identity in, 211–26
    slowing of aging in, 227–32
    see also self-help exercises
self-esteem, 157, 186, 194–99
self-help exercises, 236
    for calming down, 90
    for charisma and empathy,
        176–78
    for confidence, 199, 200–201
    for creativity, 207–10
    for depression, 105–6, 108–10,
        111–13
    for developing new identity,
        217–23
    for friendship, 172–73
    for guilt trips, 99–100
    for happiness, 15–16, 25–26,
        28–29
    for healthy diet and exercise,
        129–30
    for long-term relationships, 69–70
    for love games, 66–67
    for moving on, 64–65
    for persuasion, 162–64, 168–69
    in posture for perseverance, 138
    for procrastination, 114, 126–27
    to slow effects of aging, 231–32
    ten most effective, 243–45
    for weight control, 132, 137
    for willpower, 141
Seligman, Martin, 83
separate-but-equal doctrine, 156–57
September 11, 2001, terrorist attacks,
    148–49
serotonin, 103
sex:
    Freudian approach to, 85, 92
    love and, 36
    research into, 39
Shakers, love and, 36
Shaw, George Bernard, 231
Sherif, Muzafer, 179–83, 185
shopping, 173
Singapore, National University of, 64,
    140
Singapore Management University, 208
singing, happiness created through,
    30–31
60 Minutes, 170
slimness, obesity vs., 133–37

SM (patient), 19

smiles, smiling, 177, 240–41

and depression, 107

happiness created by, 10–11, 13–17, 21, 22, 107, 137, 158, 192, 194–95

laughter vs., 27–28

Smith, Liz, 230

Smith College, 150

smoking:

attempts to end, 117–18, 123, 140, 142–43, 148, 149 50, 153–54, 174

of marijuana, 154

Snodgrass, Sara, 22

soccer (football), 88

Social Animal, The (Aronson), 186

social interaction, 23

persuasion in, 36–40

Soma puzzle experiment, 120–21

sour grapes, 170–71

speed dating, 44, 65–68

Sperry, Roger, 239–41

spinal cord, 19

injury to, 77–79

in neural network, 75–76

split-brain patients, 240–41

sports:

clothing colors in, 205

competition in, 181, 185

fan anger and aggression in, 87, 88

Stanford University, 121, 184, 211–12, 224

Stanislavski, Constantin, 15

Stewart, James, 229

Stockholm syndrome, 57

Stony Brook University, 60

Straus, Murray, 86

Sumner, William Graham, 156

Supreme Court, U.S., 156–58

"suproxin," 47, 51

Swarthmore College, 59

sweating, 47, 53

Syracuse University, 204

"Talk. They'll Listen" campaign, 153–54

Taylor, Elizabeth, 62

temperature:

anger and, 53

and friendliness, 172–73

skin, 18

Temple University, 87

testosterone, 199

Texas, University of, 54, 184

Texas A&M University, 138

therapeutic aggression movement, 85

"They Stoop to Conquer" (Riskind), 138

"Third Wave" experiment, 166–68

thought control, 158, 158

Time, 147

tomato, as "apple of love," 43

Tonight Show, 166

tonsillectomies, 80, 82

Toronto, University of, 82, 124

torture, foot-in-the-door technique in, 127

Touch and Go (jazz ensemble), 39
Townsend, John Marshall, 204
transportation, segregated, 157
twenty pieces exercise, 114, 126–27

Ultimatum (game), 225–26
unconscious mind, 83–84
unison, in bonding process, 183, 184
University Medical Center Groningen, 101
Uppsala University, 178
Urbach-Wiehte disease, 19

Van Swearingen, Jessie, 107
Vassilikos, Vassilis, 55–56
Velten, Emmett, 23–24
Veterans Administration hospital, 77
Victorian period, xi, 189
    psychology in, 3–6, 10, 58
Vietnam War, 159
Virtual Human Interaction Lab, 224
virtual reality, 224–26
visualization, 196

Wagner, Richard, 95
walking styles, emotions affected by, 22
Washington, University of, 113
Washington University, 17
Watson, John B., 91–95
Wave, The (Rhue), 168
Wegner, Daniel, 60
weight control:
    As If principle in, xi, 131–43

internal vs. external signals in, 133–37
    Pull Me-Push You technique for, 132
Welle, Die, 168
"What Is an Emotion?" (James), 75
White, Gregory, 54
"Wild and Crazy Guy" routine, 54
William James Fellow Award, 7
Williams, Lawrence, 173
willpower:
    As If principle in developing, xi, 115–43
    failure of rewards in maintaining, 117–23
    impact of small changes on, 124–30
    in weight loss, 131–43
Wiltermuth, Scott, 184
Wisconsin, University of, 38
Wiseman, Richard, xi, 54, 89
    happiness study of, 20–21
    hypnotism experiment of, 237–39
    new form of speed dating of, 65–68
Wittman, Blanche, 83–84
Wolpe, Joseph, 95
women:
    depression in, 102
    in sex research, 39
    in workplace, 215
words, language, emotions created by, 23–24, 25–26

workplace:

    habit in, 139–40

    women's identity in, 215

World of Warcraft, 224–26

World War II, 202

worm experiment, 194–96

"Would You..." (Touch and Go), 39

Wundt, William:

    ball-dropping experiment of, 3–4, 5, 6

James vs., 5–6, 12

Watson vs., 91

Yee, Nick, 224–26

yellow bile, 190

Yom Kippur, 135–36

Youngman, Henny, 148n

Zhong, Chen-Bo, 172

Zimbardo, Philip, 211–15, 227

About the Author

Richard Wiseman is based at the University of Hertfordshire in the United Kingdom and is Britain's only Professor for the Public Understanding of Psychology. He has an international reputation for his research into unusual areas, including deception, luck, humor, and the paranormal; is frequently quoted by the media; and has been featured on more than 150 television programs across the world.